P9-EDV-651

MAPLE LABS
FOR
LINEAR ALGEBRA

TERRY LAWSON
Tulane University

ROBERT J. LOPEZ
Rose-Hulman Institute of Technology

JOHN WILEY & SONS, INC.
New York • Chichester • Brisbane
Toronto • Singapore

Copyright © 1996 by John Wiley & Sons, Inc.

All rights reserved

Reproduction or translation of any part of this work
beyond that permitted by Section 107 and 108 of
the 1976 United States Copyright Act without the
permission of the copyright owner is unlawful.
Requests for permission or further information
should be addressed to the Permission Department,
John Wiley & Sons, Inc.

ISBN 0-471-13594-1

Printed in the United States of America

10 9 8 7 6 5 4 3 2 1

Printed and bound by Malloy Lithographing, Inc.

PREFACE

These labs are a transcription into Maple of an equivalent set of MATLAB linear algebra labs keyed to Terry Lawsons text, *Linear algebra*. The original set of labs was written by Terry Lawson and is accompanied by a number of additional MATLAB files that generate the LAWSON TOOLBOX.

The interesting Maple challenge was to determine how much additional functionality had to be added to Maple itself to achieve the functionality provided by the LAWSON TOOLBOX. Although the MATLAB version of this Supplement was written with the Student Version 4.0 of MATLAB in mind, and that version gives access to some of the symbolic functionality of Maple, only one additional Maple file is needed to reproduce the functionality of the LAWSON TOOLBOX. This one file, written by Dr. Eugene Johnson of the University of Iowa, is being made available by his generosity and courtesy. The file generates the command **rrefshow**, an animated version of the reduction of a matrix to reduced row echelon form (reduced normal form). The code for **rrefshow** can be downloaded by anonymous FTP from either of the two sites listed at the end of this Preface.

There is one other file that the user of the Maple version of this Supplement would profit from acquiring. This is the file *colvects* which instructs Maple to display and print its vectors as true columns. This file was written by Mike Monagan, now at Simon Fraser University in Burnaby, British Columbia, Canada. In fact, much of the structure of the present *linalg* package in Maple reflects Mike Monagans view of linear algebra. This file, and instructions for installing and using it, can be obtained, by anonymous FTP, from either of the two sites listed at the end of this Preface.

The default behavior of Maple with respect to defining vectors is to create an object that behaves as if it were a column vector, but which is displayed horizontally, as if it were to be interpreted as a row vector. These objects are not row vectors and should not be so interpreted. It is merely an optical illusion! These horizontally written objects have all the properties of column vectors.

Maple *is* biased, however, against row vectors and actually favors column vectors in certain aspects. Ordinarily, the transpose of a column vector is defined as a row vector, and the transpose of a row vector is a column vector. In Maple, applying the command **transpose** to a column vector, even one that has been displayed vertically as a column, is not visually transposed. The display merely shows the word **transpose** applied to the column vector. But such an object will behave as if it were now a row vector!

The file *colvects* has been tested with Maple V Release 2, Maple V Release 3, and Maple V Release 4. The version of Maple used in the preparation of this Supplement is Release 3. When teaching the linear algebra course at Rose-Hulman Institute of Technology, and in my own personal work, I have the colvects code installed in my Maple initialization file. Since this installation is platform-specific, instructions for how to convert the file *colvects* to a Maple initialization file are provided in the FTP subdirectory in a file called *ReadMe*. Since most computations in Maple are displayed to the screen and not printed, the space consumed by printing vectors as columns costs nothing. For the final printing of an assignment to hardcopy, the visual tricks embodied in the *colvects* file can be turned off or can be retained to make reading the students paper more intelligible.

Maple is modularized for efficiency and memory management. There are presently some 24 "packages" in Maple, where a package is a cluster of related commands pertaining to a specific area of mathematics. These packages are integral to Maple, and every correctly installed copy of Maple will have all 24 of these packages. However, we need to direct the reader into just the *linalg*, *student*, *stats*, *plots*, and *orthopoly* packages.

It is a significant task to acquire a working familiarity with large portions of Maple. There are more than 2500 commands, a resident library containing some 250 "everyday" functions, and the 24 packages. The *linalg* package alone contains more than 100 commands. It would be a daunting task to work through these labs if they were predicated on a mastery of even the full *linalg* package. A check of the Index of Maple Commands on page 105, however, shows that only 47 of the commands from the linalg package are used in this Supplement, and only 104 Maple commands in total are needed at all. Since these 104 commands are spread out over the 13 labs in this Supplement, it is unlikely that Maple syntax will be a problem when implementing the computations of linear algebra in Maple.

In addition to the two files just mentioned, we have included all the Maple inputs for each of the 13 labs in the form of 13 Maple Worksheets. A Maple Worksheet is the interactive record of a Maple session. It can be relaunched at a later time, and the commands in it re-executed, modified, and extended. Copy/paste and in-line editing work in any Worksheet run under a windowed environment. So Worksheets are a very flexible medium for interacting with mathematics through Maple. Instructions for downloading and installing these Worksheets are included in the ReadMe file at the anonymous FTP site listed above. The Worksheets are stored and transported as ASCII text files, but the actual manipulations to get them functioning are platform specific.

Since the structure of the original version of these labs is determined by Lawson's *Linear Algebra* text, we call attention to the early and continuing use of the formalism his text adopts for Gaussian elimination when solving the system $Ax = b$. Forward elimination yields the equivalent system $O_1 Ax = O_1 b$, with $O_1 A$ an upper triangular matrix U that is not necessarily unique. Hence, the matrix O_1, storing a record of the elementary row operations used in the Gaussian elimination, is likewise not unique.

If the backward eliminations are then carried out, however, the original system is transformed to the equivalent $OAx = Ab$, where $OA = R$, the reduced normal form (reduced row echelon form), *is* unique. Hence, the matrix O is unique.

The LAWSON TOOLBOX of MATLAB commands contains code for generating O_1, O, and R. Since these matrices can be obtained in Maple by applying four built-in Maple commands, it was not deemed necessary to build an equivalent operator for the Maple version of these labs. It is a trivial exercise in using the Maple programming language to build such a procedure, but building it well takes considerable care. Good Maple code includes error detection, trapping for improper input, robustness against naive users, and so on. The interested reader can surely pursue this if so inclined.

The best source for information about Maple is Maple itself. An extensive help system is included within Maple. When running Maple under any of the Windowing systems available on today's computers (Motif for UNIX platforms, Microsoft Windows for PCs, etc.) there will be pull-down menus, with keyboard equivalents, for launching the Help Browser. The Browser gives a logically arranged, tree-structured interface to all the help files. Nearly all such help files contain a section of examples illustrating the use of the command in question. The tree-structure helps to locate the actual command name being sought.

In addition to the Help Browser, there is also keyword search wherein Maple looks for all possible terms related to the keyword supplied. Finally, for those commands whose names are known, the use of the question mark (?) from the keyboard will launch the help file. For example, ?matrix executed from the keyboard will launch the help file for the Maple command **matrix**.

A working copy of Maple needs to be installed and running before the downloaded Worksheets can be executed. The Student Edition of Maple (supplied by the Brooks/Cole Publishing Company) will suffice for running these labs. The Maple code in the Student Edition is the same as in the Standard Edition, but there is a memory hobble that limits the size of the problems that can be solved. None of our labs reaches this memory limit.

Working through these labs requires executing the Maple instructions so that the outputs can be studied, and then answering the interspersed questions posed. The questions are not accumulated at the ends of the labs, but are to be found embedded in the dialog these labs aim to initiate with the student.

The typical Maple prompt is the greater-than symbol (>) for most platforms. This can be changed to a bullet (•) on a Macintosh, but we have retained the wedge for consistency. Students should not type the wedge. The prompt is provided by the Maple program itself, and the student should type only what appears in these labs after the prompts. Every Maple line must terminate with appropriate punctuation, typically a semicolon, and sometimes a colon (when output is to be suppressed.) Instructions are sent to Maple by hitting the Enter key. Some platforms distinguish between a **Return** key and an **Enter** key. A simple experiment will help a novice user determine the proper key, but there is an extended discussion of this issue in Lab 1.

The labs can be studied in conjunction with the course material as a way of making the text's discussions concrete. They can be assigned as homework, or they can be incorporated into a formal laboratory session. My experience teaching with Maple in the classroom since 1988 suggests that the more quickly the novice Maple user is steered around the potential pitfalls, the sooner Maple becomes a valuable tool for learning mathematics. Probably the worst format for these labs would be one in which the students struggle with Maple syntax as they try to deduce the mathematics, and there is no help available to remedy either kind of difficulty. Having some help available when students are doing these activities will be the most user-friendly environment.

At Rose-Hulman Institute of Technology we have been using tools like Maple in the classroom since 1988. We began using Maple for calculus instruction and then progressed to differential equations, circuits in electrical engineering, and dynamics in mechanical engineering. Classes are taught in rooms equipped with one computer for each student. For the fall quarter of 1995 each member of our incoming freshman class purchased a laptop computer to provide the computing environment needed in all courses throughout the curriculum.

Certainly it is not necessary to have such facilities available to profit from this Maple Supplement. Whatever the level of technology available, we encourage its use. With the advent of symbolic calculators in the mold of the TI-92, we believe that we are on a path whereby all our math and science courses will be included in a framework of a ubiquitous technology. This Maple Supplement attempts to provide the appropriate instruction for the student implementing linear algebra in Maple.

Finally, we provide an anonymous FTP site at which code (and instructions) for the files *colvects* and *rrefshow* can be found. In addition, this site contains, in the form of Maple Worksheets, each of the 13 labs of this Supplement.

ftp.rose-hulman.edu

See the subdirectory /pub/users/lopez/linalg

CONTENTS

LAB 1

Basic Maple Commands

Contents:

Maple Commands: *+, -, *, /, :=, ?, ;, :, augment, col, equal, evalf, evalm, extend, matrix, print, row, seq, stack, submatrix, transpose, vector, with(linalg)*

Topics: Entering matrices and vectors, linear combinations, matrix multiplication, rows and columns, scalar multiplication, submatrices

This lab introduces the basic features of Maple. After the Maple prompt (typically >) type

> with(linalg);

and press the **Enter** key to cause this first input to be executed in Maple. Note that on the PC and UNIX platforms there is just an **Enter** key, but on a Macintosh there is both an **Enter** key and a **Return** key. On the Macintosh the **Return** key does not send input to Maple for execution. Instead, it generates a new line. (The equivalent behavior on other platforms would be obtained by simultaneously pressing the **Shift** and **Enter** keys.)

Maple is subdivided into libraries and packages to provide a modularity that induces efficiency. There are some 24 packages, and *linalg*, the linear algebra package, will be our most commonly used package. The command just executed makes available some 105 new commands that are not present in the resident library.

Maple V Release 3, under any windowing environment, provides Worksheets as the document style for a Maple session. The Worksheet itself is the record of the session, and allows on-screen editing and reexecution of commands. Saving the Worksheet saves the record of the session. On a Macintosh, no special file extension is needed for Maple to recognize the file as that of a Worksheet. However, for a PC or UNIX platform, filenames for Maple Worksheets take the "ms" extension. Hence, using the File/Save menu options, give the session a name such as Lab1 on the Macintosh and Lab1.ms on other platforms. The PC version allows files to be saved and opened from icons on the toolbar. Otherwise, to relaunch a saved Worksheet, either double click on the file's icon (Macintosh) or launch Maple first and use the File/Open menu options on other platforms. The relaunched Worksheet typically does not retain any memory of previous results so all the commands would need to be reexecuted to cause the variables to be redefined. It is possible to save the Worksheet along with the actual memory state of the computer at the time of the save, but such memory files are large. Since it is generally only a moment's work to reexecute a Worksheet, saving the memory state is generally not necessary.

Murphy's Law may well have been formulated specifically for the computer environment of today's world. Save Maple sessions early and often.

Within the Maple session, commands typically end with a semicolon. Ending an input command with a colon suppresses the output. The command is executed, but the output is not printed. The command executed above loads the code for doing linear algebra and makes some 105 linear algebraic functions available to the user. After noting some of the linear algebra commands in the list echoed, one might resolve to load the package with the colon and to rely on the Maple Help Browser to search for command names and the syntax for them.

To launch the Maple Help Browser use the appropriate menu item. For example, on the Macintosh, this is found under the menu heading "Windows," while on the PC the Browser is launched either from the menu heading "Help" or by clicking the question mark on the toolbar. The Help Browser is organized as a logical tree, with each of the columns opening up to the right as the depth of the tree is penetrated. In particular, to access, in the Browser, the complete list of commands in the *linalg* package, click on Mathematics (column 1), Linear Algebra (column 2), linalg... (column 3). Column 4 will contain an alphabetical list of the *linalg* commands. Selecting one of these commands (by clicking) generates a brief description at the bottom of the Browser, and clicking the "Help" button on the Browser will launch a full Help Screen, complete with examples, for that command. Of course, if the name of the command is already known, then the Browser can be bypassed and the same Help Screen launched directly from the Worksheet by executing a Maple command of the form

> ?matrix

The principal data structures supported by Maple for use in linear algebra are the matrix and the vector. Although both matrices and vectors can be created by an **array** command, these labs will distinguish between the matrix and vector and will not lump both of these under the heading of "array."

Vectors are not matrices, and they are not lists. They are created with the command **vector** as follows.

> vector([1,2,3]);

The default response by Maple will be to echo the vector "horizontally," with entries separated by spaces. This object behaves as a *column* vector, even though it is printed on the screen horizontally. The file *colvects*, containing Maple code for forcing Maple to print these vectors as columns, can be obtained, by anonymous FTP, from either of the two sites listed at the end of the Preface. This file can either be read into each Maple session or installed in a Maple initialization file, thereby changing the default behavior of Maple "permanently." Dr. Michael Monagan of Simon Fraser University in Burnaby, British Columbia, Canada, is the author of this code.

In its native state, Maple will write a vector in a horizontal mode but will treat the object as if it were a true column vector. Maple does not allow a row vector to be input directly. There is no data structure in Maple corresponding to a row vector. To obtain the equivalent of a row vector, one must transpose a column vector. The transpose of an array is the object in which columns have become rows and rows have become columns. For a column vector, the transpose is then a "horizontal object" and is called a "row vector." The syntax for creating a row vector in Maple will be illustrated shortly.

Care should be exercised in distinguishing vectors from lists. Entering

> [1, 2, 3];

creates a list which will be echoed back by Maple with the entries separated by commas, not spaces.

The assignment operator in Maple is := and it is used to attach a name to a Maple object. Thus

> v := [1,2,3];

creates a list whose name is "v" and

> V := vector(v);

creates the vector whose name is V. Obviously, Maple is case sensitive since it distinguishes between upper and lower case letters. Spaces are ignored by Maple and are included here only to make the text easier to read. Note, however, that a space between the colon and the equal sign in the assignment operator will generate a syntax error.

For scalar quantities, Maple echoes the contents of a variable when interrogated for it with the following syntax.

> x := 3;
> x;

Having assigned the variable x the value 3, we can ask Maple to print this value for us by entering the variable name. However, this is not the case with vectors and matrices. Asking Maple to display the contents of the variable V, which we know is a vector, yields just the symbol "V." To get Maple to display the contents of both vectors and matrices, use, for example the **print** command or the **evalm** command.

> print(V);
> evalm(V);

Assignments are "erased" by assigning the variable name its literal (or "letter") value via the single forward quotes. The assignment

> x := 'x';

restores the symbol x to its status as an unassigned variable.

Maple behaves as if the transpose of its vectors, by default "column" vectors, are row vectors. However, Maple only echoes to the screen a grudging reference that it has understood a request to transpose a vector. Thus,

> Vt := transpose(V);

creates a row vector Vt, but Maple only responds with the echo *transpose(V)*. Later, we'll see ways to determine Maple's actual behavior with these objects. If code has been installed whereby Maple prints its column vectors as actual columns, then the expectation is that the transpose of a column vector will be printed as a row. Again, note that Maple only echoes the phrase *transpose(V)*.

The kth component of the vector V can be referenced, and even changed, through the notation V[k] as in the following example.

> print(V);
> V[2] := t;
> print(V);

The matrix is most easily entered with the **matrix** command whose syntax is illustrated below.

> A := matrix(2,3, [1,2,3,x,y,z]);

Including the dimensions of the intended matrix allows Maple to wrap the list of elements into the correct rows. Alternatively, the matrix can be entered as a "list of lists" with each sublist being one of the rows of the matrix.

> matrix([[1,2,3], [x,y,z]]);

Moreover, the syntax of the **matrix** command is flexible enough to allow the entries of the matrix to be generated by a user-specified function. Consider, for example,

> matrix(2,3, (i,j) -> x^(i + j));

in which the "arrow" -> is created by typing a minus sign and a greater-than sign. The arrow is one of the ways Maple allows functions to be entered, and the syntax just illustrated implements a function that maps the integer pairs (i,j), (where i = 1, 2, and j = 1, 2, 3), to the symbols x^{i+j}.

The matrix $\begin{pmatrix} a11 & a12 \\ a21 & a22 \end{pmatrix}$ can be entered in Maple by means of the concatenation operator "." that appends symbols such as "a" and "1" or "2".

> matrix(2,2, (i,j) -> a.i.j);

Individual entries in a matrix can be addressed and changed via the syntax

```
> A[1,2] := t;
> print(A);
```

Matrices can also be assembled from vectors. Keep in mind that to Maple a vector is a "column-thing." The transpose of a vector, while tolerated by Maple, is not a suitable building block from which to build a matrix. Applying the following **stack** and **augment** commands to the transpose of a vector results in a syntax error. In the following illustration of these commands we also create the vector **v2** whose entries are the symbols x1, x2, and x3. Note the Maple syntax that creates the sequence x1,x2,x3. Inside the square brackets, this sequence becomes a list, the proper argument for the **vector** command.

```
> v1 := vector([1,2,3]);
> v2 := vector([x.(1..3)]);
> ?augment
> ?stack
> A := augment(v1, v2);
> B := stack(v1,v2);
```

Augment builds a 2x3 matrix whose columns are the (column) vectors **v1** and **v2**, whereas **stack** builds a 3x2 matrix whose rows are the elements of **v1** and **v2**. If **stack** or **augment** are applied to a single vector, a *MATRIX* results. It is important to observe this distinction, since the vector data structure is changed by these commands. It might appear typographically that applying **augment** to a vector has produced a column vector, but note carefully that the output of **augment** is a *MATRIX*, not a vector. In Maple these are different objects. Applying **stack** to a vector again produces a *MATRIX*, not a row vector. Failure to realize that Maple distinguishes between vectors and matrices can make Maple seem opaque to the naive user.

Another way to enlarge a matrix uses the **extend** command. For example

```
> ?extend
> extend(A, 2, 3, 0);
```

adds two rows and three columns of zeros to the matrix A.

Rows and columns can be extracted from a matrix by the **row** and **col** commands.

```
> r2 := row(A, 2);
```

creates the vector (necessarily a column vector) **r2** whose components are the entries in the second row of the matrix A.

```
> c2 := col(A, 2);
```

creates the vector (again, necessarily a column vector) **c2** whose components are the entries

in the second column of the matrix A.
    The **submatrix** command extracts a submatrix from an existing matrix. For example,

> submatrix(A, 2..3, 1..2);

is the submatrix formed from the second and third rows, and the first and second columns of A. It is also possible to skip rows or columns, as illustrated by the following usage.

>submatrix(A, [1,3], 1..2);

Question 1. (a) Enter the 3x4 matrix $A = \begin{pmatrix} 2 & 2 & -1 & 3 \\ 1 & 4 & -2 & 1 \\ 3 & 2 & 0 & 9 \end{pmatrix}$.

(b) Assign its (3,4) entry to the variable *A34*.
(c) Assign its third row to the variable *row3*.
(d) Assign its second column to the variable *col2*.
(e) Form a matrix B which has as its rows the second and third rows of A.
(f) Form a Matrix C which has as its columns the first, second, and third columns of A.
(g) Form a matrix F whose entries come from the first and third rows of A, and the third and fourth columns of A.

(h) Form a matrix G whose columns are the fourth and first columns of A in the order A^4, A^1.

Question 2. (a) Input the lists L1 = [1,3,2,1], L2 = [2,4,1,-1], and the (column) vectors

$c1 = \begin{pmatrix} 1 \\ 2 \\ 4 \end{pmatrix}$, $c2 = \begin{pmatrix} 1 \\ 1 \\ -1 \end{pmatrix}$, $c3 = \begin{pmatrix} 0 \\ -3 \\ -1 \end{pmatrix}$.

(b) Form a matrix A whose rows are the entries in L1, L2.
(c) Form a matrix B whose rows are L2 and L1, in this order.
(d) Form a matrix C whose columns are c1, c2, c3.

(e) Without changing the name of C, add to it as a fourth column, the vector $\begin{pmatrix} 1 \\ 2 \\ 3 \end{pmatrix}$.

(f) Change the (2,4)-entry of C to 0.
(g) Replace the third row of C by [4,3,2,1].

    We now look at matrix multiplication.

> A := matrix(3,3, [seq(k, k = 1..9)]);
> B := matrix([[4,2], [1,3], [4,1]]);

We have used Maple's **seq** operator to generate the sequence of integers 1, 2, ..., 9. The matrix product of A and B can be obtained in Maple in two ways.

```
> multiply(A, B);
> evalm(A &* B);
```

The first usage can be extended to several matrices but is still not as flexible as the second. The second usage allows any amount of matrix arithmetic, but at the expense of needing the **evalm** (evaluate-matrix) command. Note that noncommutative multiplication in Maple requires the &* operator. Use of * either will give a syntax error or will run the risk of having Maple commute the factors.

Multiplication of a vector v by the matrix A can be accomplished by either technique, but in general, **evalm** proves to be the more desirable approach.

```
> v := vector([2, -1, 1]);
> multiply(A, v);
> evalm(A*v);
```

Entering the product as v*A yields an answer in Maple, but it is easily seen that Maple has commuted the factors to yield a dimensionally correct product, namely, A*v.

```
> evalm(v*A);
```

However, if the noncommutative operator &* is used, then

```
> evalm(v &* A);
```

generates a syntax error because left-multiplication of a matrix by a (column) vector is not defined, and Maple cannot now commute the factors. For the same reasons, the product AB requires &* and BA, even with &*, is not possible.

```
> evalm(A &* B);
> evalm(B &* A);
```

To create the row vector $w = (2 \ -1 \ 1)$ in Maple, enter a column vector and apply the **transpose** command. Remember, the transpose will not show on the screen as a row vector.

```
> w := transpose(v);
```

The product **w**A is dimensionally correct, so in Maple it can be computed by

```
> evalm(w*A);
```

because that is the only correct order for the factors. Maple's output is a row vector, which is expressed as the transpose of a column vector.

Multiplication of the square matrix A by itself, say, three times, is accomplished by

```
> evalm(A^3);
```

Arithmetic operations on matrices and vectors are most easily performed in Maple with the **evalm** command. It is possible to avoid using **evalm**, but this is at the expense of a more tedious use of individual commands for matrix addition and subtraction, and for scalar multiplication. Thus, avoiding **evalm** requires that multiplication use **multiply**, scalar multiplication use **scalarmul**, addition use **add**, and subtraction use **add** with additional parameters. We recommend the use of **evalm**. Matrices of different sizes can't be added, and a request for Maple to try generates an error message.

```
> A := matrix([ [3,2,1,4], [3,2,1,5], [4,2,3,5] ]);
> B := matrix([ [2,1,1,1], [-2,1,-1,0], [2,1,3,2] ]);
> C := matrix(4,2, [seq(k, k = 1..8)]);
> evalm(A + B);
> evalm(A - B);
> evalm(A + C);
> evalm(5*A);
> evalm(5*A - 3*B);
```

Question 3. Use the matrices A, B, C given above, and the (column) vectors **v** and **w** whose respective entries are (1,3,-1,2) and (-1,2,1,4). Form the following linear combinations and products. Warning: Some of these may be impossible; try to predict these cases before typing them in.

(a) 3A
(b) 5A + 3B
(c) 3A + 4B
(d) **v** + **w**, A**v**, A**w**, A(**v** + **w**), A**v** + A**w**
(e) 5**v**, A(5**v**), 5A**v**
(f) AC, BC, (A + B)C, AC + BC
(g) CA

Unlike other numeric programs, Maple performs its calculations symbolically, not in floating point arithmetic. Thus, symbols like x and y, and symbols like 1/2, $\sqrt{2}$, and π are all treated as symbols in Maple. Arithmetic is therefore exact and suffers no round-off error. Of course, Maple converts to floating point representation upon user demand, and this is done with the **evalf** (evaluate-floating) command. Moreover, since Maple has the ability to manipulate as many as 500,000 digits, it is, for all practical purposes, an arbitrary precision calculator. Try, for example, the following calculations.

```
> evalf(sqrt(2));
> evalf(sqrt(2), 25);
> evalf(Pi);
> evalf(Pi, 5000);
```

> 1/2 + 1/2;
> 1/2 + .5;

Question 4. Enter the 5x5 matrix A whose (i,j) element is $a_{i,j} = \dfrac{i-j}{i+j}$ and compute (using Maple's default exact arithmetic) A^{10}.

Question 5. Create A, B, and C, three completely symbolic 3x3 matrices with entries of the form a11, a12, and so on. Use these matrices to check the associative law (AB)C = A(BC). To test for equality of two matrices P and Q in Maple, use the command equal(P, Q) or evaluate the difference P - Q.

LAB 2

2. Gaussian Elimination

Contents:

Maple Commands: *addrow, augment, diag, evalm, ffgausselim, gausselim, linsolve, matrix, mulrow, rref, submatrix, subs, vector, with(linalg)*

Supplementary routines: *rrefshow*

Topics: Forward elimination, Gaussian elimination, permutations, reduced normal form, row operations

The fundamental problem in the first chapter is solving the equation Ax = b by Gaussian elimination. Maple has built-in routines that implement this technique for solving linear systems. The basic Maple solver for algebraic equations is the **solve** command. In the event the equations to be solved are linear and in matrix form, then the **linsolve** command in the *linalg* package is available. When the solution is not unique, **linsolve** includes arbitrary parameters in its solution. The meaning of these parameters is discussed in Chapter 1 of the text <u>Linear Algebra,</u> by Terry Lawson. We can illustrate the application of the **linsolve** command with the following example in which the matrix A and the vector **v** is entered. The vector **b** = A**v** is calculated and the system A**x** = **b** is solved. The system does not have a unique solution, so **x** ≠ **v**.

```
> with(linalg):
> A := matrix(3,3, [seq(k, k = 1..9)]);
> v := vector([4,3,2]);
> b := evalm(A*v);
> x := linsolve(A, b);
> evalm(A*x);
```

We see that the solution vector **x** contains arbitrary parameters, indicating that the solution is not unique. Hence, A**x** = **b**, but **x** is not **v**.

Question 1. Explain why **x** contains arbitrary parameters, what these parameters tell us, and why $\mathbf{x} \neq \mathbf{v}$.

On the other hand, if we use linsolve to seek a solution to the equation $A\mathbf{x} = \mathbf{c}$, where $\mathbf{c}$ is the vector $\begin{pmatrix} 1 \\ 0 \\ 0 \end{pmatrix}$, we find from

```
> c := vector([1,0,0]);
> linsolve(A, c);
```

that there is just a null return from Maple, indicating there is no solution to this system of equations.

The vector $\mathbf{x}$ that linsolve finds as solution to $A\mathbf{x} = \mathbf{b}$ is the general solution containing the arbitrary parameter $\_t_1$. It is instructive to write the solution as $\mathbf{x_0} + \_t_1\mathbf{x_1}$, where $\mathbf{x_0}$ is a particular solution and $\mathbf{x_1}$ spans the null space of A. However, the immediate concern here is Gaussian elimination itself. We will perform these computations on the augmented matrix

```
> Ab := augment(A, b);
```

We wish to eliminate the (2,1)-, (3,1)-, and (3,2)-elements of Ab by Gaussian elimination. This is implemented in Maple with the **addrow** command which adds r times row i to row j (in matrix M) via the syntax addrow(M, i, j, r). First note, however, that not all texts order the indices in the same way. For example, in the text Linear Algebra by Terry Lawson, the notation O(j,i;r) is used to indicate that r times the row i is added to row j. Thus,

```
> A1 := addrow(Ab, 1, 2, -4);
> A2 := addrow(A1, 1, 3, -7);
> A3 := addrow(A2, 2, 3, -2);
```

The matrix Ab has been lower-triangularized, leading to a solution of the system $A\mathbf{x} = \mathbf{b}$. To recover this solution, multiply the second equation in A3 by -1/3, then use the 1 so created to eliminate the (1,2)-element of the resulting matrix. Note the use of the **mulrow** command which multiplies a specific row by a designated scalar.

```
> A4 := mulrow(A3, 2, -1/3);
> A5 := addrow(A4, 2, 1, -2);
```

Question 2. Explain how to obtain the solution vector $\mathbf{x}$ from matrix A5.

Question 3. Augment A with the vector $\mathbf{c} = \begin{pmatrix} 1 \\ 0 \\ 0 \end{pmatrix}$ and row reduce by Gaussian elimination as was done with Ab. Explain why this will show that there is no solution to the system $Ax = c$.

Given the augmented matrix Ab, we next use Maple to reduce it to normal form (reduced row echelon form), extract the reduced forms of A and **b**, reconstitute the reduced equations, solve these equations for the basic variables, and generate the general solution in terms of the particular solution and the free variables.

```
> Abr := rref(Ab);
> Ar := submatrix(Abr, 1..2, 1..3);
> br := subvector(Abr, 1..2, 4);
> x := vector([x.(1..3)]);
> q1 := evalm(Ar*x);
> q2 := student[equate](q1,br);
> q3 := solve(q2, {x1, x2});
> xx := subs(q3, evalm(x));
```

The vector **xx** will contain the sum of the particular solution and the appropriate multiples of the free variables. These can easily be sorted by inspection. Sorting in Maple requires careful attention to the delicate evaluation rules associated with the **evalm** command. First, the particular solution **xp** is extracted by setting $x3 = 0$. Next, the homogeneous solution is extracted by subtracting **xp** from **xx**. Then, a scaled basis vector is obtained from **xh** by setting $x3 = 1$, and finally, the vector **xx** is reassembled with x3, **xh1**, and **xp** all distinctly displayed.

```
> xp := subs(x3 = 0, evalm(xx));
> xh := evalm(xx - xp);
> xh1 := subs(x3 = 1, evalm(xh));
> x3*evalm(xh1) + evalm(xp);
```

Bringing a matrix into reduced normal form requires more work than bringing the matrix into echelon form (the result of doing only the forward eliminations.) Although echelon form is not unique (while the reduced normal form is), the basic and free variables are just as evident in the echelon form as in the reduced normal form. Maple nearly provides the echelon form with its **gausselim** command; the leading entries in each nonzero row may not be 1. Since this is easily remedied with **mulrow**, we consider reduction to echelon form (forward elimination) to be resolved with built-in Maple commands. For example,

```
> fe := gausselim(Ab);
> fe1 := mulrow(fe, 2, -1/3);
```

does forward elimination on the augmented matrix Ab. The steps used to write the solution as a sum of a particular solution and multiples of the free variables will work on the matrix fe1.

We again caution the reader that row reducing a matrix to upper triangular form is not a unique process. When Maple does this reduction by its **gausselim** command, the computations are done the same way each time, so in that sense the result is "unique." But if the reader were to do the reduction by hand, or stepwise with the **addrow** command in Maple, unless the same steps were used as in the **gausselim** algorithm, the results would not necessarily be the same. This could be disconcerting to the reader trying to match a computation done by hand with the output of **gausselim**. Hence, it behooves us to point out just what steps **gausselim** implements.

Maple's **gausselim** command does not interchange rows unless a zero appears on the main diagonal. Since Gaussian elimination requires division by the diagonal elements, such a zero divisor presents a problem. Typically, one would interchange rows to move a nonzero divisor into place. In such a case Maple looks for the nearest row with a nonzero divisor. Since any row below the row with the potential zero divisor can be so exchanged, we have discovered the source of the non-uniqueness in the process of upper triangularizing a matrix. So, any reader attempting to reproduce, by hand, the output of **gausselim**, need only be sure to interchange no rows unless there is a zero divisor on the diagonal. In this case, interchange the row containing the zero divisor with the first row beneath it having a nonzero divisor.

Question 4. For the system $A\mathbf{x} = \mathbf{b}$, with A and $\mathbf{b}$ given by

$$A = \begin{pmatrix} 1 & 2 & 3 & 4 \\ 5 & 6 & 7 & 8 \\ 9 & 10 & 11 & 12 \end{pmatrix}, \mathbf{b} = \begin{pmatrix} 10 \\ 26 \\ 42 \end{pmatrix},$$

show that the matrix resulting from forward elimination yields the general solution as a sum of a particular solution and multiples of the free variables by the technique detailed above.

Then use forward elimination to show that the system $A\mathbf{x} = \mathbf{b_2}$, where $\mathbf{b_2} = \begin{pmatrix} 1 \\ 0 \\ 0 \end{pmatrix}$, has no

solution.

A movie form of **gausselim** is available in the supplementary Maple code called **rrefshow**, available by anonymous FTP at either of the two sites listed at the end of the Preface. The author of the code is Dr. Eugene Johnson of the University of Iowa.

Terry Lawson's <u>Linear Algebra</u> introduces two symbols, O_1 and O, for keeping track of the Gauss elimination steps performed on a matrix A. The matrix O_1 records those steps used to bring A to an upper triangular form. (See, for example, Sections 1.5 and 1.8.) Since this upper triangular form is not unique, the matrix O_1 recording the transition is not unique. However, if we compute O_1 by means of the **gausselim** command (or its equivalent) then we will always get the same result. Recall the discussion just after Example 3 where we discussed this issue at length.

Linear Algebra speaks of the reduced normal form R for a matrix. This unique form is sometimes called the reduced row echelon form (rref) in other texts. The matrix recording the steps that bring A into the unique form R is called O in Lawson's text. Hence, $O_1 A =$ U and OA = R express the reductions in the notation used in Lawson. Once again, since the canonical form R is unique, the algorithm by which it is calculated is immaterial. Both R and the matrix O that transforms A to R are unique. It is only the matrix O_1 that transforms A to U that we have cautioned the reader about.

The matrix O can be obtained in Maple using the **rref** command which transforms A completely to its reduced normal form (sometimes called reduced row echelon form). These distinctions are illustrated in the following computations. In each case, A is augmented by an identity matrix of appropriate size, and the augmented matrix is reduced by Gaussian elimination. Depending on the extent of the reductions, we can obtain either an O_1 matrix or the O matrix. Either of these matrices will be the result of performing the same Gaussian steps on the identity as were performed on A itself. The matrix products $O_1 A$ and OA are then the same as the reduced versions of A, respectively U and R, since both O_1 and O "record" the Gaussian elimination steps performed on A. And finally, one last time, the Maple implementation of Gaussian elimination embodied in Maple's **gausselim** command is consistent with the algorithm used in Lawson's text for obtaining both O_1 and U. It is only the reader, doing the reduction by hand and then wishing to compare the result with results generated by Maple, who needs to be concerned about the possible non-uniqueness of O_1 and U.

```
> A := matrix(3,3, [8,4,-5,-5,3,-5,8,5,-1]);
> id := diag(1$3);
> A1 := augment(A, id);
> A2 := gausselim(A1);
> A3 := ffgausselim(A1);
> A4 := rref(A1);
> O1 := submatrix(A2, 1..3, 4..6);
> U1 := submatrix(A2, 1..3, 1..3);
> O2 := submatrix(A3, 1..3, 4..6);
> U2 := submatrix(A3, 1..3, 1..3);
> o := submatrix(A4, 1..3, 4..6);
> evalm(O1 &* A);
> evalm(O2 &* A);
> evalm(o &* A);
```

The identity matrix id is created with the **diag** command that takes a sequence of three 1's (created by 1$3) and puts these 1's on the main diagonal of id. Both **gausselim** and **ffgausselim** reduce A to upper triangular forms. The second is done without introducing fractions, while the first requires rational arithmetic and results in the introduction of fractions. The transformed identity that results from **gausselim** is O_1 while the transformed identity from **ffgausselim** is O_2, and these two matrices are not the same. Moreover, the reduced forms of A, namely, the matrices U1 and U2, are not the same either. The third matrix, o, is unique, since it records the transformation of A to its unique reduced normal form, which for this matrix A, is the identity. As much as we might want to call this matrix O, Maple uses the character O for another purpose and will not let us assign to it. A warning is issued and we are forced to adopt another symbol.

In each case the products O_1A, O_2A, and OA all reproduce the reduced form of A, the differing upper triangular matrices possible from Gauss elimination. In the last case we obtained R, the reduced normal form itself.

Question 5. For the matrix A given above and the vector $\mathbf{b} = \begin{pmatrix} 4 \\ 1 \\ 5 \end{pmatrix}$, reduce the system Ax = b to $U_1\mathbf{x} = O_1\mathbf{b}$ and also to $U_2\mathbf{x} = O_2\mathbf{b}$. Show that the resulting solutions for **x** are the same.

Question 6. For the matrix A of Question 4, use the vector $\mathbf{b} = \begin{pmatrix} b_1 \\ b_2 \\ b_3 \end{pmatrix}$ and the gausselim command to determine the condition on b necessary for solutions of Ax = b to exist. Show further that the conditions on **b** generated by the **ffgausselim** command are equivalent.

LAB 3

Invertibility, Transposes and Determinants

Contents:

Maple Commands:	*addcol, addrow, augment, col, det, diag, dotprod, equal, evalm, expand, gausselim, inverse, linsolve, matrix, minor, nullspace, plot, rank, rref, seq, simplify, tsubmatrix, ranspose, vector, with(linalg)*
Supplementary routines:	*rrefshow*
Topics:	Column operations, determinant, Gauss-Jordan method, invertibility, rank, transpose

This lab studies invertibility, transposes, and determinants. First, we determine when a matrix is invertible.

```
> with(linalg):
> A := matrix([ [1, -1, 2], [0, 1, 3], [-2, 1, 1] ]);
```

The rank of a matrix A is the number of nonzero rows in the reduced normal form; we find it using

```
> rank(A);
```

If the matrix is invertible, it must have rank equal to the number of rows. The Gauss-Jordan method of finding the inverse augments the matrix with the identity and seeks, by Gaussian elimination, to reduce the matrix itself to the identity. If the submatrix on the left successfully reduces to the identity, then the submatrix on the right has been transformed into the inverse. If the Gaussian elimination fails because a row of zeros develops on the left, then the original matrix is not invertible. Implementing this technique for finding an inverse requires that we generate an identity matrix of appropriate dimensions. Although it seems natural to assign the identity to the symbol I, in Maple this letter is reserved for $\sqrt{-1}$. Hence, we will use id for the matrix identity.

```
> id := diag(1$3);
> AI := augment(A, id);
```

The first row operation is

```
> A1 := addrow(AI, 1, 3, 2);
```

Question 1. Continue performing row operations until the lefthand side is the identity. Maple has the command **inverse** which computes the inverse of a matrix. Check your answer using

```
> inverse(A);
```

You can also check your result by applying Maple's **rref** command to AI. If A in AI is reduced to the identity, then the identity in AI will have been transformed into A^{-1}.

Here are two more examples to work through. Start by doing the operations on the matrix B to try to reduce it to the identity. If you can get B to the identity, then perform the same operations on the identity matrix to get the inverse of B. Check your answers with **inverse** and **rank**.

Question 2. (a) $B = \begin{pmatrix} 1 & 1 & 1 \\ 1 & 2 & 1 \\ 1 & 1 & 2 \end{pmatrix}$; (b) $B = \begin{pmatrix} 1 & 1 & 1 \\ 1 & 2 & 1 \\ 1 & 0 & 1 \end{pmatrix}$.

Question 3. For the 2x2 matrix $C = \begin{pmatrix} a & b \\ c & d \end{pmatrix}$, show that the inverse exists only if ad - bc = det(C) is nonzero.

The transpose of a matrix is the new matrix formed by interchanging the rows with the columns. Hence, row 1 becomes column 1, column 1 becomes row 1, and so on. If the general element in A is $a_{i,j}$, then the general element in the transpose, A^t, is $a_{j,i}$. The Maple command for obtaining a transpose is **transpose**. By way of example, we illustrate, for 2x2 matrices, the rule that the transpose of a product is the product of the transposes in the reverse order: $(AB)^t = B^t A^t$.

```
> A := matrix(2,2, (i,j) -> a.i.j);
> B := matrix(2,2, (i,j) -> b.i.j);
> AB := evalm(A &* B);
> C1 := transpose(AB);
> C2 := evalm(transpose(B) &* transpose(A));
> equal(C1, C2);
```

By inspection we see that $C1 = (AB)^t = B^tA^t = C2$.

The solvability of the system $A\mathbf{x} = \mathbf{b}$ hinges on $\mathbf{b}$'s being completely in the column space of A. If multiplication by the m by n matrix A is thought of as a mapping from R^n to R^m, then the partition of R^n into the column space of A^t and its orthogonal complement, the null space of A, and the partition of R^m into the column space of A and the null space of A^t determine those vectors $\mathbf{b}$ for which the equation $A\mathbf{x} = \mathbf{b}$ has a solution. If $\mathbf{b}$ lies completely in the column space of A, then $A\mathbf{x} = \mathbf{b}$ can be solved. This means $\mathbf{b}$ must have no component in the null space of A^t, a condition equivalent to requiring $\mathbf{b}$ to be orthogonal to every vector in $N(A^t)$. To begin studying these relationships, augment A with a general vector $\mathbf{b}$ to form Ab, and perform Gaussian elimination on Ab. Solvability conditions on $\mathbf{b}$ will be contained in any rows in Ab where A has been reduced to zeros but $\mathbf{b}$ has not.

```
> A := matrix([ [2,1,0,2,1], [-2,0,2,-3,-1], [-2,1,4,-4,-1] ]);
> b := vector([b.(1..3)]);
> Ab := augment(A, b);
> Abr := gausselim(Ab);
```

The bottom row of Abr dictates that $-b_1 - 2b_2 + b_3 = 0$. This condition that states $\mathbf{b}$ must be orthogonal to the null space of A^t. In Maple, we can find a basis for this nullspace as follows.

```
> C := nullspace(transpose(A));
> c := C[1];
```

Maple returns C, a set containing the single basis vector $\mathbf{c} = \begin{pmatrix} -1 \\ -2 \\ 1 \end{pmatrix}$. The construct C[1] selects the first (and here, the only) vector contained in this set. The condition on $\mathbf{b}$ is equivalent to requiring $\mathbf{b}$ to be orthogonal to $\mathbf{c}$.

```
> dotprod(b, c) = 0;
```

Finally, we can check that $A^t\mathbf{c} = 0$.

```
> evalm(transpose(A) * c);
```

An alternative route to finding the null space of A^t would use **linsolve** for solving the equations $A^t\mathbf{x} = 0$ for $\mathbf{x}$. Let $\mathbf{z5}$ be the zero-vector with five components. As we have seen for the construction of identity matrices, the Maple syntax 0$5 will create a sequence of five zeros.

```
> z5 := vector([0$5]);
> linsolve(transpose(A), z5);
```

The solution indicates that any multiple of the vector **c** is in $N(A^t)$.

If desirable, we can explore the way Gauss elimination generates the solvability condition on **b**. If we augment A with a 3x3 identity and subject the augmented matrix to the **rref** command, the part of the augmented matrix where the identity was now contains the matrix O as defined by Lawson's text. Hence, Ax = b can be transformed to OAx = Ob, and the product Ob generates the same condition on **b** that we found above. Note again that Maple will not allow assignment to the symbol O since that already has a defined meaning. We use "o" instead.

```
> id := diag(1$3);
> AI := augment(A, id);
> AI1 := rref(AI);
> o := submatrix(AI1, 1..3, 6..8);
> evalm(o*b);
```

Question 4. Find all solutions of Ax = 0 by (a) using Maple's **nullspace** command applied to A and (b) using **linsolve**. Show that both methods produce the same null space. Note that the **nullspace** command finds a basis for the null space, while **linsolve** finds the general solution of Ax = 0.

Since A is 3x5, we have m = 3 and n = 5. The column space of A^t lies in $R^n = R^5$ and is orthogonal to N(A), the null space of A. A general member of the column space of A^t is

$$A^t y, \text{ where } y = \begin{pmatrix} y_1 \\ y_2 \\ y_3 \end{pmatrix} = y_1 \begin{pmatrix} 1 \\ 0 \\ 0 \end{pmatrix} + y_2 \begin{pmatrix} 0 \\ 1 \\ 0 \end{pmatrix} + y_3 \begin{pmatrix} 0 \\ 0 \\ 1 \end{pmatrix} = \sum_{k=1}^{3} y_k e_k. \text{ If each basis vector } \mathbf{w} \text{ in}$$

N(A), found in Question 4, is orthogonal to each vector $A^t e_k$, then N(A) is the orthogonal complement of the column space of A^t. For each vector **w** in the basis of the null space of A, compute the dot product of w with $A_t e_k$, where e_k is the kth column of the 3x3 identity matrix id. Each of the nine dot products should be zero.

```
> w := nullspace(A);
> id := diag(1$3);
> seq(seq(dotprod(w[i], transpose(A)*col(id, k)), k = 1..3), i = 1..3);
```

All nine dot products are zero. The nested **seq** operator allows iteration over the three basis vectors in N(A) and over the three basis vectors e_k in R^3. The output is a sequence of nine zeros. Observe that $A^t e_k$ is the kth column of A^t and the kth row of A. The dot product of **w** with $A_t e_k$ is just one of the components of A**x**. Since the vectors **w** are solutions of the equation A**x** = **0**, the dot products computed above cannot fail to be all zero.

The determinant of a matrix A is found by det(A).

> A := matrix([[1, -1, 2], [0, 1, 2], [8, 1, 2]]);
> det(A);

Question 5. Reduce A to an upper triangular matrix U, using **addrow** and **mulrow** to implement the Gaussian elimination steps. Explain how this is related to det(A). Repeat

this for B = $\begin{pmatrix} 2 & 1 & 0 \\ 4 & 2 & 1 \\ 1 & 1 & 1 \end{pmatrix}$.

Question 6. (a) Use Maple's **det** command to obtain the determinants of A = $\begin{pmatrix} 1 & 2 & 3 & 4 \\ 5 & 6 & 7 & 8 \\ 9 & 10 & 11 & 12 \\ 13 & 14 & 15 & 16 \end{pmatrix}$ and B = $\begin{pmatrix} 1 & 0 & -1 & 1 \\ 2 & 1 & 1 & 2 \\ -1 & 0 & -1 & -1 \\ 1 & 2 & 2 & -1 \end{pmatrix}$.

(b) Use Maple's **minor** command to obtain these two determinants via a Lagrange expansion of a row or column. In most texts, a cofactor is a signed minor, and a minor is a *determinant* of the submatrix formed by deleting a row and column. However, both Terry Lawson's <u>Linear Algebra</u> and Maple take a minor to be the submatrix, not its determinant.
(c) Obtain the determinant of B by first using column 4 to create zeros in column 1, then doing a Lagrange expansion by the transformed first column. In Maple, a column operation can be performed by the command **addcol** which has syntax comparable to that of **addrow**.

(d) Use the values found in (a) to determine the determinants of (i) BA, (ii) B^2, (iii) the inverse of B, (iv) -2B.
(e) Find the determinant of C = B - xI, where x is a scalar and I is the 4x4 identity matrix. Note that det(C) is a polynomial in x. Plot this polynomial for $-3 \le x \le 3$. Illustration of the Maple **plot** comand is given below.

> ?plot
> f := x^2;
> plot(f, x = -2..2);
> plot(f, x = -2..2, y = 0..5);

The extant version of Maple, Maple V Release 3, requires that a plot, which opens in a separate plot window, be copied and pasted into the Maple Worksheet. Use the edit/copy and edit/paste menu options, first on the plot window to copy, and then on the worksheet to paste.

Question 6 examines some of the properties of determinants. In particular, we demonstrate for general 3x3 matrices that (a) the determinant of the product is the product of the determinants, (b) the determinant of a scalar multiple is more than just the scalar multiple of the determinant, and (c) the determinant of the transpose is the same as the determinant of the original matrix.

```
> A := matrix(3,3, (i,j) -> a.i.j);
> B := matrix(3,3, (i,j) -> b.i.j);
> q1 := det(A &* B);
> q2 := expand(det(A) * det(B));
> simplify(q1 - q2);
> det(x*A);
> det(A);
> det(transpose(A));
```

LAB 4

Basic Vector Space Concepts

Contents:

Maple Commands: *augment, colspace, det, diag, evalm, gausselim,*
linsolve, matrix, nullspace, rank, rowspace, rref, stack,
submatrix, transpose, vector, with(linalg)

Topics: Basis, linear combinations, linear independence, span,
row space, column space

In this lab we will use Maple to explore some connections between the concepts of
linear combinations, span, linear independence, and bases.

A fundamental notion is that of a linear combination of a collection of vectors. If
$v_1, \ldots, v_k$ is a collection of vectors, then

$$\mathbf{v} = c_1\mathbf{v}_1 + \cdots + c_k\mathbf{v}_k$$

is called a linear combination of these vectors. In Maple, the command **vector** creates an
object that behaves like a column vector. To represent a row vector in Maple, **transpose**
must be applied to a column vector. Column vectors can be made to look like columns
with the addition of special code (see Lab 1), but the transpose of a column vector is
always echoed in Maple as the operator *transpose* applied to the column vector, and is not
displayed as a row vector.

Consider the following three vectors in R^4, and the matrix B, whose columns are the
given vectors.

```
> with(linalg):
> w1 := vector([1, 2, -1, 0]);
> w2 := vector([-3, 0, 2, 1]);
> w3 := vector([2, 9, 1, -4]);
> B := augment(w1, w2, w3);
```

The vectors **w1, w2,** and **w3** are treated in Maple as if they were indeed column vectors.
In Maple, to create three row vectors with the same components requires

```
> v1 := transpose(w1);
> v2 := transpose(w2);
```

> v3 := transpose(w3);

   To form the matrix A whose rows are the three row vectors **v1**, **v2**, and **v3** requires that we apply the **stack** command to the column vectors **w1**, **w2**, and **w3**. Applying either **augment** or **stack** to row vectors gives a syntax error in Maple. Hence, we again see that Maple views the fundamental nature of a vector to be a column vector, and a row vector is merely the transpose of a column vector. There seems to be a built-in bias toward column vectors since there are operations on column vectors that have no counterpart among Maple's "row vectors."

> A := stack(w1, w2, w3);

   The linear combination $\sum_{k=1}^{3} c_k \mathbf{v}_k$ can be obtained, as a row vector, by writing the row vector $\mathbf{c} = (c_1, c_2, c_3)$ and forming the product **c**A.

> c := transpose(vector([c1, c2, c3]));
> evalm(c * A);

The product **c**A is just the linear combination of the rows of A, with coefficients coming from the entries in the row vector **c**. For example, if we want the linear combination $-2\mathbf{v}_1 + 3\mathbf{v}_2 + 5\mathbf{v}_3$, we have the options

> evalm(-2*v1 + 3*v2 + 5*v3);
> cc := transpose(vector([-2, 3, 5]));
> evalm(cc * A);

   On the other hand, working with column vectors, the equivalent linear combinations would be $\sum_{k=1}^{3} c_k \mathbf{w}_k$. In Maple, the two options for the specific numeric case above would be direct computation and the matrix product **B**d, where **d** is the column vector $\mathbf{cc}^t$.

> evalm(-2*w1 + 3*w2 + 5*w3);
> d := transpose(cc);
> evalm(B * d);

Clearly, the product **B**d is a linear combination of the columns of B with coefficients coming from the entries of **d**.

Question 1. (a) Using row vectors, find the following linear combinations, doing this both directly and by using a matrix wherein the rows are the given vectors.
(a1)   -2(2,1,0,1) + (3,4,1,2) - 7(2,1,1,1)

(a2)   4(2,1,2,4,3) + 5(9,2,1,2,2) - 3(2,1,3,2,6) - (3,4,1,1,1)
(a3) -2(2,1,2,4,3) + 3(9,2,1,2,2) - 2(2,1,3,2,6) - (3,4,1,1,1)
(b) Repeat part (a) using column vectors instead of row vectors and using a matrix whose columns are the given vectors.

Recall that the subspace spanned by the vectors $v_1, \ldots, v_k$ is the set of all linear combinations of these vectors. If we write these vectors as the columns of a matrix A, then the product Ac is a linear combination of the columns of A, hence, $c_1 v_1 + \cdots + c_k v_k$. Thus, a vector **w** is in the subspace spanned by $v_1, \ldots, v_k$ iff it is in the column space of A, that is, iff there is a solution **c** of Ac = **w**. To implement this study in Maple, start with the column vectors

```
> v1 := vector([1, 0, 2, -1]);
> v2 := vector([2, 0, 1, 2]);
> v3 := vector([-1, 1, 1, 1]);
```

and the matrix A whose columns are the given vectors.

```
> A := augment(v1, v2, v3);
```

Consider the vector

```
> w := vector([-1, 1, 4, -3]);
```

which we want to write as a linear combination of v_1, v_2, v_3. We must attempt to solve the equations Ac = **w** for some vector **c**.

```
> linsolve(A, w);
```

A null response to this command indicates there is no solution for **c**, and hence, **w** is not a linear combination of v_1, v_2, v_3. A single solution would indicate a unique representation of **w** in terms of the v_k's and a solution with parameters would indicate that there are multiple linear combinations of the v_k's that would yield **w**.

Question 2. Determine whether the following vectors are linear combinations of v_1, v_2, v_3. If they are, determine *all* values of **c** so that the given vector is expressible as $c_1 v_1 + c_2 v_2 + c_3 v_3$.
(a) w = (1,0,0,0)
(b) w = (4,0,5,0)

Question 3. Let $v_1 = (1,2,1)$, $v_2 = (0,1,2)$, $v_3 = (-1,0,3)$. Determine whether the following vectors are linear combinations of v_1, v_2, v_3. If they are, determine *all* values of **c** so that the given vector is expressible as $c_1 v_1 + c_2 v_2 + c_3 v_3$.

(a) (1,0,0)
(b) (0,3,6)

Applying the **gausselim** command to the matrix A augmented with an appropriate form of the identity matrix, we can obtain a condition on exactly which vectors **w** are linear combinations of the columns of A. The matrix that replaces the identity after the Gauss elimination steps are completed is O_1. Remember, O_1 is not unique since it depends on the manner in which the Gaussian elimination was carried out. However, the product $O_1\mathbf{w}$ captures all the Gaussian elimination steps of the reduction of A to the companion upper triangular form U, so this product is equivalent to augmenting A with **w** and performing Gauss reduction on this augmented matrix. Hence, the last m - r entries of $O_1\mathbf{w}$ have to be zero if there are m rows in A and its rank is r.

For the A in the discussion preceding Question 2, $A = \begin{pmatrix} 1 & 2 & -1 \\ 0 & 0 & 1 \\ 2 & 1 & 1 \\ -1 & 2 & 1 \end{pmatrix}$, and we can find

$O_1\mathbf{w}$ by the following Maple calculations.

```
> A := matrix([ [1,2,-1], [0,0,1], [2,1,1], [-1,2,1] ]);
> w := vector([-1,1,4,-3]);
> rank(A);
> id := diag(1$4);
> AI := augment(A, id);
> AI1 := gausselim(AI);
> U := submatrix(AI1, 1..4, 1..3);
> O1 := submatrix(AI1, 1..4, 4..7);
> evalm(O1*w);
```

We thereby find that the rank of A is 3, that U has its last m - r = 4 - 3 = 1 row completely zero, and that $O_1\mathbf{w}$ also has its last entry zero. Hence, **w** is in the column space of A and the equation A**x** = **w** has a solution.

Alternatively, augmenting A with **w** and performing Gaussian elimination are implemented by the following Maple steps. It is no surprise that entirely equivalent results are obtained.

```
> Aw := augment(A,w);
> gausselim(Aw);
```

Conditions under which A**x** = **y** has a solution for an arbitrary vector **y** can be found from the product $O_1\mathbf{y}$. Implemented in Maple, the computation

```
> y := vector([y.(1..4)]);
> evalm(O1*y);
```

produces the condition $-\frac{5}{3}y1 - 4y2 + \frac{4}{3}y3 + y4 = 0$.

To see if the following two specific vectors **Y1** and **Y2** are compatible with this condition, examine $O_1\mathbf{Y1}$ and $O_1\mathbf{Y2}$.

```
> Y1 := vector([1,0,0,0]);
> Y2 := vector([4,0,5,0]);
> evalm(O1*Y1);
> evalm(O1*Y2);
```

Question 4. Determine equations, linear in the coefficients of **w**, that must be satisfied if **w** is to be in the subspace spanned by (1,2,-1,2), (1,-1,0,1), (1,2,1,1).

To determine if the vectors $\mathbf{v_1}, \ldots, \mathbf{v_k}$ are independent, we have to see whether the equation $c_1\mathbf{v_1} + \cdots + c_k\mathbf{v_k} = \mathbf{0}$ has any solutions other than just $\mathbf{c} = \mathbf{0}$. If there is no other solution, then the vectors are independent. If there is another solution **c**, called a nontrivial solution, then the vectors are dependent. The vector **c** then gives a dependency relation between the vectors, allowing us to solve for one of the vectors in terms of the others. Using Maple to form a matrix A with columns taken as the column vectors $\mathbf{v_1}, \ldots, \mathbf{v_k}$, we

seek solutions of the equation $A\mathbf{c} = \mathbf{0}$. For example, consider the three R^3 vectors

```
> v1 := vector([1,2,-1]);
> v2 := vector([2,1,1]);
> v3 := vector([1,1,1]);
```

and the associated matrix

```
> A = augment(v1, v2, v3);
```

The zero vector in R^3 is

```
> z3 := vector([0$3]);
```

and the equation $A\mathbf{c} = \mathbf{0}$ is solved by

```
> linsolve(A, z3);
```

The response is $\mathbf{c} = \mathbf{0}$, indicating there is no nontrivial solution; the vectors are independent. However, note the linear dependence of the following column vectors.

```
> v1 := vector([2,1,0]);
> v2 := vector([1,2,1]);
> v3 := vector([0,3,2]);
```

```
> A := augment(v1, v2, v3);
> c := linsolve(A, z3);
```

This time, **c** is any multiple of the nontrivial vector (1, -2, 1), giving a dependency relation $v_1 - 2v_2 + v_3 = 0$. We can solve for v_3 as $v_3 = -v_1 + 2v_2$, so that the subspace spanned by the three vectors is actually spanned by just the first two. Vectors v_1 and v_2 form a basis for the subspace.

Question 5. Determine whether the following vectors are independent. If they are dependent, write one of the vectors as a linear combination of the others.
(a) (1,2,3,4), (5,6,7,8), (9,10,11,12)
(b) (1,0,1), (1,1,0), (1,2,-1)
(c) (1,0,0,-1), (1,0,-1,0), (1,-1, 0,0), (-3,1,1,1)

Determining whether the vectors $v_1, \ldots, v_k$ are linearly independent requires solving the equation $Ac = 0$. Hence, we are, in effect, looking for the null space of A. Maple has a built-in command for finding a basis of the null space. Maple neither normalizes nor orthogonalizes the basis vectors so found. Procedures for converting a collection of vectors to an equivalent set of orthonormal vectors will be considered in Lab 7.
As an example of a null space computation, consider the matrix

```
> B := matrix([ [2,1,0], [1,2,3], [0,1,2] ]);
> nullspace(B);
```

To determine whether a collection of vectors is a basis for a vector space (or one of its subspaces), we must know that they span the vector space and are also independent. Two subspaces of seminal importance for interpreting a matrix are the row space and the column space. The row space is that space spanned by the rows of the matrix, while the column space is that space spanned by the columns of the matrix. For example, consider the matrix

$$A = \begin{pmatrix} 1 & 2 & 3 \\ -1 & 2 & 1 \\ 2 & -1 & 2 \\ -1 & 1 & 0 \end{pmatrix}$$ for which the row space is a subspace of R^3 and the column space, a

subspace of R^4. The span of the rows of A is certainly a subspace in R^3, but before we can declare these rows a basis we need to know if they are linearly independent. One way of determining the independence of the rows is to examine the reduced normal form of A.

```
> A := matrix([ [1,2,3], [-1,2,1], [2,-1,2], [-1,1,0] ]);
> R := rref(A);
```

The nonzero rows of R form a basis for the row space so the dimension of the row space is given by the number of nonzero rows of R. (For additional discussion of this point, see Proposition 2.2.1 or p.132 of Lawson's text.)

Since the rows of A form a spanning set for the row space, a basis can be extracted from them. If R = rref(A) has r nonzero rows, then there are r rows in A that form a basis, and any collection of more than r rows from A must be linearly independent.

Finally, it is instructive to compare the nonzero rows of R with the output of Maple's built-in **rowspace** command. The same vectors are produced!

```
> rowspace(A);
```

For this example, there are three independent rows in A. The row space has dimension 3, so the standard basis for R^3 is realized as the basis for the row space. This dimension of the row space of A is called its **row rank**. If we define the dimension of the column space as the **column rank**, we can prove that these two dimensions are always the same number, and this common number is then called the **rank** of A. Maple has the built-in commands

```
> rank(A);
> colspace(A);
```

Finding a basis for the column space again starts with Gaussian elimination. However, the row operations of this process are not linear combinations of the *columns* of A. Hence, one cannot merely select the nonzero columns of the reduced normal form of A as a basis for the column space of A. What does turn out to be true, however, is that the number of basic columns (columns with basic variables) is the column rank of A, and the columns of A itself that correspond to the positions of the basic columns in the reduced normal form of A do indeed form a basis for the column space of A.

Question 6. For the matrix $B = \begin{pmatrix} 1 & 2 & 3 & 4 \\ 5 & 6 & 7 & 8 \\ 9 & 10 & 11 & 12 \end{pmatrix}$, find bases for the row space and the column space.

It is instructive to compare the basis of the column space suggested by the discussion above, with the basis generated by Maple's **colspace** command. In fact, this command row-reduces B^t, a reduction that forms linear combinations of the rows of B^t, and hence, of the columns of B itself. For the matrix B of Question 6 we find

```
> B := matrix(3,4, [seq(k, k = 1..12)]);
> colspace(B);
> rowspace(transpose(B));
```

Having discussed finding bases for the row and column spaces of A, we next consider finding a basis for the null space of A. However, we have already shown that such a basis arises from the solution of $B\mathbf{x} = \mathbf{0}$. Moreover, the Maple command **nullspace** also yields a basis for the null space.

Finally, we consider finding a basis for the null space of B^t. Of course, we can merely transpose B and apply to B^t either of the techniques mentioned above. There is one other way to look at the null space of B^t. Augmenting B with an appropriate identity matrix and then row reducing by Gaussian elimination leads to the matrix O_1 for which the product O_1B is the upper triangular matrix U that is the outcome of the Gaussian elimination on B. Hence, we have the equation $O_1B = U$. Now for B, an m by n matrix, U has its last m - r rows all zeros. Moreover, the product of the matrices XY can be viewed as taking each row of X and dotting it with each column of Y. If that view is taken in the product O_1B, then each of the last m - r rows of O_1 must be orthogonal to the columns of B. The columns of B define the column space of B, and the null space of B^t is the orthogonal complement of the column space of B. Consequently, the last m - r rows of O_1 must be in the null space of B^t.

Question 7. For the matrix B of Question 6, find O_1 and U. Show that there are m - r rows of zeros at the bottom of U. Show that the bottom m - r rows of O_1 satisfy $B^ty = 0$. Use the **nullspace** command to find a basis for the null space of B^t and compare the vectors obtained by the two methods.

A set of exactly n vectors in R^n will be a basis for R^n if the vectors span R^n or if they are linearly independent. A test for independence consists of creating a matrix A whose columns are the given set of vectors, and showing that the null space of A contains just the zero vector. Equivalently, the vectors are independent if the rank of A is n. For example, the vectors (1,2,1), (2,1,3), (2,4,1) can be shown to be a basis for R^3 by the Maple computations below.

```
> v1 := vector([1,2,1]);
> v2 := vector([2,1,3]);
> v3 := vector([2,4,1]);
> A := augment(v1, v2, v3);
> nullspace(A);
> rank(A);
> det(A);
```

If the null space of A contains just the zero vector, or if the rank of A is 3, or if the determinant of A is nonzero, we can conclude that the vectors v_1, v_2, v_3 are linearly independent. Should any one of these three conditions fail to hold, then we would conclude that the vectors were not independent, as the following example demonstrates.

```
> v1 := vector([1,2,1]);
> v2 := vector([2,1,1]);
> v3 := vector([3,3,2]);
```

```
> A := augment(v1, v2, v3);
> nullspace(A);
> rank(A);
> det(A);
```

Question 8. In each case, determine whether the following sets of vectors form a basis for R^4. If they do not form a basis, find a basis for the subspace which they span.
(a) (1,0,0,0), (1,1,0,0), (1,1,1,0), (1,1,1,1)
(b) (1,2,3,4), (5,6,7,8), (9,10,11,12), (13,14,15,16)
(c) (1,0,1,-1), (-1,1,0,1), (-1,0,1,-1), (1,-1,-2,1)
(d) (1,0,1,0), (0,1,0,1), (1,0,-1,0), (0,1,0,-1)

LAB 5

Linear Transformations in the Plane

Contents:

Maple Commands: *display, evalm, matrix, plot, seq, vector, with(linalg), with(plots)*

Topics: Plotting polygons and circles, and plotting their images under linear transformations

Linear transformations from R^n to R^m correspond to multiplication by an m by n matrix A. We will look at the case $m = n = 2$ and use Maple and its graphics to study these linear transformations. We assume Maple V Release 3 throughout. There may be some subtle platform-specific differences in how you tell Maple which printer to address. You will have to consult your local network expert if you have difficulty printing any of the graphs in this Lab.

We will create some geometric objects and see what happens to them when we multiply by certain matrices. Since a linear transformation is determined by its values on a basis, we first choose a triangle with vertices $\mathbf{0}, \mathbf{e_1}, \mathbf{e_2}$. The linear transformation L_A that corresponds to multiplication by the matrix A will then send this triangle to one with vertices $\mathbf{0}, A\mathbf{e_1} = A^1, A\mathbf{e_2} = A^2$.

We first draw this triangle. Anticipating the need to distinguish visually between the initial and transformed triangles, we include specific options in the plot command so that the triangles will have distinctive characteristics when plotted. We first show how to use thickend lines to distinguish between the triangles. Eventually we show how to use color, in addition to thickening, to make these distinctions. We could use different line styles (solid versus dashed), but where such lines coincide the distinctions are lost. Other options to the **plot** command cause equal scaling to be used on the axes and cause the axesto be removed from the plot. All such effects except coloring can be achieved interactively in the plot window as well, but interactive application of plot options cannot be permanently attached to the plot data structure. The plots are assigned to variables so that they can later be superimposed by the **display** command which is accessed by loading the *plots* package. Options included in the **plot** command remain with the plot data structure and are preserved by the **display** command.

> with(linalg):

```
> with(plots):
> p1 := vector([0,0]);
> p2 := vector([1,0]);
> p3 := vector([0,1]);
> P := [p1,p2,p3,p1];
> f := plot(P, scaling = constrained, axes = none, thickness = 3):
> f;
```

Color is added to the original triangle by altering the **plot** command to

```
> f := plot(P, scaling = constrained, axes = none, thickness = 3, color = red):
> f;
```

Of course, on a monochrome screen the use of the color option is ineffective. Likewise, printing a colored graph on a printer that prints only with black ink is similarly ineffective.
    The determined reader can explore all the Maple plot options through the on-line help files accessed via

```
> ?plot[options]
```

    We want to see where the original triangle is sent by the linear transformations defined by the following matrices. Let

$$A = \begin{pmatrix} 3 & 0 \\ 0 & 2 \end{pmatrix}, B = \begin{pmatrix} 1 & 0 \\ 0 & -1 \end{pmatrix}, C = \begin{pmatrix} 0 & -1 \\ 1 & 0 \end{pmatrix}$$

$$F = \begin{pmatrix} \frac{1}{\sqrt{2}} & \frac{-1}{\sqrt{2}} \\ \frac{1}{\sqrt{2}} & \frac{1}{\sqrt{2}} \end{pmatrix}, G = \begin{pmatrix} \frac{1}{\sqrt{2}} & \frac{1}{\sqrt{2}} \\ \frac{1}{\sqrt{2}} & \frac{-1}{\sqrt{2}} \end{pmatrix}$$

Enter the matrix A, apply it to the vertices of the triangle P, then plot it against the graph of triangle P. These steps are done in Maple via the following commands.

```
> A := matrix(2,2, [3,0,0,2]);
> PA := [seq(evalm(A*P[k]), k = 1..4)];
> fA := plot(PA, scaling = constrained, axes = none):
> fA;
> display([f, fA], view = [-3..3, -3..3]);
```

Question 1. Using the matrices B, C, F, and G, repeat the procedures used for matrix A. The view option on the **display** commands will guarantee that the absolute size of each image retains equivalent proportions.

    Multiplication by each of the other matrices B, C, F, and G admits a simple geometric interpretation. Multiplication by B gives a reflection in the x-axis. Multiplication by B sends e_1 to itself, but sends e_2 to its negative. It will send any vector (x,y) to (x,-y), which is what the reflection does. Multiplication by C sends the vector e_1 to e_2 and the vector e_2 to $-e_1$. This is just a rotation by an angle of $\pi/2$, or 90 degrees.

Question 2. Multiplication by F corresponds to a rotation by an angle θ. By inspection, determine the angle through which the triangle is rotated. Then verify that multiplication by F is a rotation through that angle by checking how much each of the basis vectors e_1 and e_2 are rotated. This is sufficient since a linear transformation is determined by its values on a basis.

Question 3. Multiplication by G corresponds to a reflection through a line having slope $\tan(\theta_G)$, and passing through the origin. Find the angle θ_G by noting where this reflection sends each of the vectors e_1 and e_2: $Ge_1 = (\frac{1}{\sqrt{2}}, \frac{1}{\sqrt{2}})$, $Ge_2 = (\frac{1}{\sqrt{2}}, \frac{-1}{\sqrt{2}})$. Thus, the x-axis is mapped to the line $\theta = \pi/4$, and the y-axis, to the line $\theta = -\pi/4$, by this reflection. Note that when you reflect through $\theta = \theta_G$, θ_G is the average value of the angles of an initial line and the reflected line. Verify your answer for θ_G by forming the vector $(\cos(\theta_G), \sin(\theta_G))$. Multiplication by G should send this vector to itself. This also gives a trial and error method for finding θ_G by experimentally computing the image of

$$G\begin{pmatrix} \cos(\theta_G) \\ \sin(\theta_G) \end{pmatrix} \text{ for various guesses for } \theta_G.$$

Instead of looking at this triangle, we could choose other triangles and look at their images. For example, we could look at the triangle with vertices (1,0), (-1,1), (0,-1). To draw this triangle at the same scaling as the original triangle, use a similar Maple strategy.

```
> pa1 := vector([1,0]);
> pa2 := vector([-1,1]);
> pa3 := vector([0,-1]);
> Pa := [pa1, pa2, pa3, pa1];
> fa := plot(Pa, scaling = constrained, axes = none):
> fa;
> display([f, fa], view = [-3..3, -3..3]);
```

Question 4. For the matrices F and G above, draw plots that show the images of the original and new triangle on the same graph.

Question 5. Draw a plot which shows, on the same graph, the first triangle and its image under B.

Besides triangles, we can look at other polygonal curves. For example, we can form a square and look at its image. Note that the original square is drawn with thickend lines.

```
> v1 := vector([1,0]);
> v2 := vector([0,1]);
> v3 := vector([-1,0]);
> v4 := vector([0,-1]);
> V := [v1,v2,v3,v4,v1];
> fv := plot(V, scaling = constrained, axes = none, thickness = 3):
> VA := [seq(evalm(A*V[k]), k = 1..5)];
> fvA := plot(VA, scaling = constrained, axes = none):
> display([fv, fvA], view = [-3..3, -3..3]);
```

Question 6. For each of F and G, draw the square and its image on the same graph.

Question 7. Draw a pentagon with the given vertices, and draw its image under the reflection G, if the vertices are $(1,0)$, $(\cos(2\pi/5), \sin(2\pi/5))$, $(\cos(4\pi/5),\sin(4\pi/5))$, $(\cos(6\pi/5),\sin(6\pi/5))$, $\cos(8\pi/5),\sin(8\pi/5))$.

For a properly installed copy of Maple connected to its own printer or to a printer on a network, printing should be no problem. Use the menu option for printing, found under the File heading.

Besides looking at graphs of polygonal curves and their images, we can also look at smooth curves such as circles. Although a Maple plot consists of a polygonal line connecting a sequence of points, the details need not concern the user of the **plot** command. A transformation such as that connected with matrix A represents a change of coordinates: $\begin{pmatrix} x' \\ y' \end{pmatrix} = A\begin{pmatrix} x \\ y \end{pmatrix} = A\begin{pmatrix} \cos(\theta) \\ \sin(\theta) \end{pmatrix}$ for a point constrained to lie on the unit circle.

Hence, to plot the image of a circle under the action of the matrix A, we need only implement the obvious mathematics in Maple, noting that both the circle and its image are given parametrically with parameter t. The Maple **plot** syntax conforms to this recognition.

```
> X := vector([cos(t), sin(t)]);
> f1 := plot([X[1], X[2], t = 0..2*Pi], thickness = 3):
> q := evalm(A*X);
> f2 := plot([q[1], q[2], t = 0..2*Pi], scaling = constrained):
> display([f1, f2], scaling = constrained);
```

This shows that A sends the unit circle into an ellipse.

Question 8. From your knowledge of the effect of each of the linear transformations L_A, L_B, L_C, L_F, L_G, predict what the image of the unit circle will be under their actions. Explain your predictions. Draw the image of the unit circle under each of these linear transformations to verify your predictions.

We now look at some more general 2x2 matrices, and use Maple to explore what happens when we multiply by them. These are somewhat special in that they are symmetric matrices.

$$F_1 = \begin{pmatrix} 2 & 1 \\ 1 & 2 \end{pmatrix}, F_2 = \begin{pmatrix} -1 & 1 \\ 1 & -1 \end{pmatrix}$$

Question 9. For each of F_1 and F_2, plot the original triangle we looked at and plot its image under multiplication by the matrix. Do the same for the unit circle and its image. By looking at the results, find a nonzero vector v so that $F_1 v$ lies on the same line as v. Do the same for F_2. Note: These will be different vectors for different matrices. They come from the axes of the image ellipses. Check your answer by computing $F_1 v$, and so on. There are nontrivial vectors w with $F_2 w = 0$. Explain how that is related to the images of the triangle and circle under F_2. Find such a vector.

LAB 6

Matrix Representatives for a Linear Transformation

Contents:

Maple Commands:	*augment, collect, colspace, det, evalm, inverse, linsolve, matrix, nullspace, rank, rowspace, rref, solve, student[equate], transpose, vector, with(linalg)*
Topics:	Bases for fundamental subspaces, change of basis formulas, coefficients with respect to a basis, matrix representing a linear transformation, transition matrices

Maple provides an abundance of tools for obtaining bases for the four fundamental subspaces of a matrix representing a linear transformation.

(1) N(A) - A basis for the null space of A can be found by nullspace(A), linsolve(A,**0**), or rref(A). The third option requires that the rows of rref(A) be interpreted as equations defining solutions for the basic variables.

(2) R(A) - A basis for the range (column space) of A can be found by colspace(A) or by rref(A). In the latter case, the basis consists of the columns of A that correspond to the basic columns of rref(A).

(3) $R(A^t)$ - A basis for the row space of A (range of A^t) can be found by rowspace(A), colspace(transpose(A)), or rref(A). In the third option, the nonzero rows of rref(A) are a basis for $R(A^t)$.

(4) $N(A^t)$ - A basis for the null space of A^t can be found by nullspace(transpose(A)), and from rref(transpose(A)) if its rows are interpreted as equations for the basic variables. Also, as explained in Lab 5, the last m - r rows (for $A^{m \times n}$) of O_1, found by augmenting A with an appropriate identity matrix and row reducing by gausselim, form a basis for $N(A^t)$.

Question 1. Determine bases for the four fundamental subspaces of the matrix A =

$$\begin{pmatrix} 1 & 2 & 3 & 4 \\ 5 & 6 & 7 & 8 \\ 9 & 10 & 11 & 12 \end{pmatrix}.$$ Use at least two different methods for finding each basis. If the two
methods give different bases, show that the different bases are equivalent by expressing
each vector in one basis as a linear combination of the vectors in the other basis.

Suppose we know one basis $v_1, \ldots, v_k$ for a vector space V but have another candidate,
$w_1, \ldots, w_k$, vying for the role of basis. This alternate basis must have exactly k vectors,
and need only be shown to be independent or a spanning set. Equivalently, if we can
express each member of the alternative basis in terms of the original basis, then the
candidate will also be a basis. Since $v_1, \ldots, v_k$ form a basis, each vector w_j must be a
linear combination of these basis vectors, so that $w_j = t_{1j}v_1 + t_{2j}v_2 + \cdots + t_{kj}v_k$ for each j
= 1,..., k. The coefficients $t_{i,j}$ form a matrix T which must therefore be nonsingular
(invertible). Note that the coefficients $t_{i,j}$ for w_j form the jth column of the matrix T. The
invertibility of T guarantees that we could express the original basis vectors in terms of the
new basis vectors with a matrix $S = T^{-1}$.

We know that 1, x, x^2 is a basis for $P^2(R,R)$, the polynomials of degree less than or
equal to 2. To check whether the polynomials $1 - x + x^2$, $2 + 2x + x^2$, $-x + 2x^2$ also form a
basis, we write these new polynomials in terms of the basis 1, x, x^2. The coefficients of
each polynomial in the new basis form the columns of the matrix T. Thus, T can be
entered row-wise and then transposed, or T can be formed from vectors of coefficients.

```
> with(linalg):
> v1 := vector([1,-1,1]);
> v2 := vector([2,2,1]);
> v3 := vector([0,-1,2]);
> T := augment(v1, v2, v3);
```

The invertibility of T can be checked in a variety of ways.

```
> det(T);
> rank(T);
> inverse(T);
> rref(T);
```

These checks should tell you that these three polynomials do, in fact, form a basis and that
the matrix which expresses the standard basis in terms of the new basis is given by the
matrix T^{-1}, computed in Maple by inverse(T).

We can also verify that these three polynomials form a basis by noting they are
independent. There is a theorem which states that whenever a collection of vectors (here
polynomials) has the same number of elements as the dimension of the space, they will

form a basis if they either span the space or are independent. A test of independence requires solving for $\mathbf{c} = (c_1, c_2, c_3)$, the equation

$$c_1(1 - x + x^2) + c_2(2 + 2x + x^2) + c_3(-x + 2x^2) = 0.$$

Rewriting this, we get

$$(c_1 + 2c_2)1 + (-c_1 + 2c_2 - c_3)x + (c_1 + c_2 + 2c_3)x^2 = 0.$$

The only way that a polynomial can be identically 0 is for all the coefficients to be themselves identically zero. Hence, we have three equations in three unknowns, namely, those captured in $\mathbf{Tc} = \mathbf{0}$. Consequently, independence is equivalent to the invertibility of T.

Question 2. Determine whether the polynomials $1, 1 + x, 1 + x + x^2$ form a basis for $P^2(R,R)$.

One way to deal with a vector space other than R^n is to use the isomorphism between it and R^n that arises when the coefficients of vectors in a fixed basis in the vector space are mapped to vectors in R^n. Then, questions about vectors in the original space can be translated into equivalent questions about the corresponding vectors in R^n. For example, in Question 2, we can map the three given polynomials to the R^3 vectors $(1,0,0)$, $(1,1,0)$, $(1,1,1)$, and ask if these triples form a basis for R^3.

Question 3. Rework Question 2 from the perspective of isomorphism with R^3.

When there are two bases V and W for a vector space S, then there is a transition matrix T_w^v whose columns describe how to write the vectors in V in terms of the vectors in W. The most direct way to determine the transition matrix begins with a study of how vectors are transformed from the V basis to the standard basis. We use the following example to make this clear.

Example 1. Start with basis vectors $\mathbf{v_1}$, $\mathbf{v_2}$, and $\mathbf{v_3}$, given in terms of the standard basis vectors $\mathbf{i}, \mathbf{j}$, and $\mathbf{k}$. If we accept the isomorphism between an arbitrary vector space and R^n then we can see the fundamental role that the standard basis plays in R^n. We will then take

an arbitrary vector $\mathbf{X_v} = \begin{pmatrix} a \\ b \\ c \end{pmatrix}_v = a\,\mathbf{v_1} + b\,\mathbf{v_2} + c\,\mathbf{v_3}$ given in terms of the basis V, and find its representation as $X_s = \alpha\,\mathbf{i} + \beta\,\mathbf{j} + \gamma\,\mathbf{k}$ in the standard basis. For example, consider the following.

```
> v1 := i + 2*j + k;
> v2 := i + k;
> v3 := -i + j + k;
> Xv := a*v1 + b*v2 + c*v3;
> Xs := collect(Xv, [i, j, k]);
```

The result of the **collect** command is $(a+b-c)\mathbf{i} + (2a+c)\mathbf{j} + (a+b+c)\mathbf{k}$, careful inspection of which shows that if the coefficients of the basis vectors $\mathbf{v_1}$, $\mathbf{v_2}$, and $\mathbf{v_3}$ are placed in a matrix V as columns, then multiplication of the column vector $\mathbf{X_v}$ by V yields the vector $\mathbf{X_s}$ whose entries are the coordinates of the vector in the standard basis. Hence, we have discovered how to relate vectors given in the basis V to their counterpart given in the standard basis S. We express this relationship as $\mathbf{X_s} = T_s^v\,\mathbf{X_v}$. Clearly, this implies $\mathbf{X_v} = (T_s^v)^{-1}\mathbf{X_s} = T_v^s\mathbf{X_s}$, so we also have determined how to map a vector given in the standard basis S to its counterpart in the basis V.

Before we continue with this example, we reinforce the notation. If the vectors in the basis V are given in terms of the standard basis S, and we "drop" the vectors $\mathbf{v_1}, \ldots, \mathbf{v_n}$ into a matrix T as its columns, then we call this matrix T_s^v, the transition matrix from V to S. Note the way the superscript "v" appears "high" on the T, poised to "drop" into the matrix as a column.

Now assume that there is another basis W consisting of the vectors $\mathbf{w_1}$, $\mathbf{w_2}$, $\mathbf{w_3}$ that are also given in terms of the standard basis S. There is a transition matrix W relating vectors in the W and S bases. In particular, we have $\mathbf{X_s} = T_s^w\,\mathbf{X_w}$ and $\mathbf{X_w} = (T_s^w)^{-1}\mathbf{X_s} = T_w^s\mathbf{X_s}$.

We are now in a position to ask how vectors given in the V basis are related to their counterparts in the W basis. If we start with the vector $\mathbf{X_v}$ we know that $\mathbf{X_v} = T_v^s\,\mathbf{X_s}$ and that $\mathbf{X_s} = T_s^w\,\mathbf{X_w}$. Consequently, $\mathbf{X_v} = T_v^s T_s^w\,\mathbf{X_w} = T_v^w\mathbf{X_w}$ and $\mathbf{X_w} = (T_v^w)^{-1}\mathbf{X_v} = T_w^v\mathbf{X_v}$. The transition matrix from V to W is therefore $T_w^v = (T_s^w)^{-1}(T_v^s)^{-1} = (T_s^w)^{-1}T_s^v$.

Example 2. Let $\mathbf{v1} = (1,2,1)$, $\mathbf{v2} = (1,0,2)$ be a basis for V, a subspace of R^3, and let $\mathbf{w1} = (1,4,0)$, $\mathbf{w2} = (0,2,-1)$ be a second basis for this same subspace. The notation implies that the basis vectors have been described in terms of S, the standard basis in R^3. Obtain the transition matrix T_w^v, the matrix that takes vectors given in the V basis to their counterparts given in the W basis. There is a special difficulty here because we are working with bases for a subspace. The transition matrices T_s^v and T_s^w that we might be tempted to write would both be 2x3 matrices and therefore not invertible. Our device of mapping a vector back to its representation in the standard basis S cannot be easily implemented here. We adopt, instead, a direct approach based on solving the equations a w_1 + b w_2 = v_1 and c w_1 + d w_2 = v_2 for the coefficients a, b, c, and d. If we write matrices V2 and W2 with columns {v_1, v_2} and {w_1, w_2} respectively, and write T = $\begin{pmatrix} a & c \\ b & d \end{pmatrix}$, then the defining equations are captured in the matrix equation W2 T = V2.

(Maple reserves the symbol W as the name of a special function. We are thereby forced to use names like V2 and W2, the 2 associating these matrixes with Example 2.) This system yields to Maple's **linsolve** command.

```
> v1 := vector([1,2,1]);
> v2 := vector([1,0,2]);
> w1 := vector([1,4,0]);
> w2 := vector([0,2,-1]);
> V2 := augment(v1, v2);
> W2 := augment(w1, w2);
> T := linsolve(W2, V2);
```

Example 3. Let V = {1, 1+x, 1+x+x^2} be a basis for P^2(R,R), the space of polynomials of degree n ≤ 2, and let W = {2-x, x+x^2, 1+x^2} be a second basis for the same space. To find the transition matrix T_w^v, we implement the isomorphism existing between P^2(R,R) and R^3 by replacing these polynomials with their coefficients with respect to the standard basis S = {1, x, x^2}.

```
> v1 := vector([1,0,0]);
> v2 := vector([1,1,0]);
> v3 := vector([1,1,1]);
> w1 := vector([2,-1,0]);
> w2 := vector([0,1,1]);
> w3 := vector([1,0,1]);
> TVs := augment(v1,v2,v3);
> TWs := augment(w1,w2,w3);
> TVw := evalm(inverse(TWs) &* TVs);
```

The Maple definitions of the vectors v_1, v_2, v_3, and w_1, w_2, w_3, set up a change of basis problem in R^3. The matrices TVs and TWs represent the transition matrices T_s^v and T_s^w respectively, where S is the standard basis in R^3. Since we are not dealing with subspaces of R^3 each matrix is nonsingular and hence invertible. The calculation of T_w^v as TVw encounters no such difficulties as faced in Example 2.

Question 4. Find the transition matrix T_w^v in the following cases.
(a) V = {(1,2,3), (2,-1,2), (2,1,4)}, W = {(5,1,9), (1,3,5), (1,0,1)}.

(b) $V = \{(0,-9,5,5), (5,18,0,10)\}$, $W = \{(1,0,2,-4), (2,9,-1,3)\}$.

(c) $V = \{1 + 3x + x^2, 3 - x + 3x^2\}$, $W = \{1 + x + x_2, 1 - x + x^2\}$.

We next consider the matrix that represents, in a given basis, the action of a linear transformation, and how this matrix is affected by a change of basis in the underlying vector space. The basic result for such a matrix is captured in the formula $[L]_W^W = T_W^V[L]_V^V T_V^W$.

Example 4. Consider the linear transformation L which is reflection in the line y = x. The basis $V = \{v1 = (1,1), v2 = (-1,1)\}$ conforms neatly to this linear transformation since $L(v1) = v1$, $L(v2) = -v2$. This means that $[L]_V^V = \begin{pmatrix} 1 & 0 \\ 0 & -1 \end{pmatrix}$. If the second basis S =

$\{e_1, e_2\}$ is the standard basis, then there is the transition matrix $T = T_S^V = \begin{pmatrix} 1 & -1 \\ 1 & 1 \end{pmatrix}$, and T_V^S

is its inverse. Thus, we compute

```
> v1 := vector([1,1]);
> v2 := vector([-1,1]);
> TVs := augment(v1, v2);
> Lvv := matrix([ [1,0], [0,-1] ]);
> Lss := evalm(TVs &* Lvv &* inverse(TVs));
```

The matrix Lss is the matrix representing the reflection with respect to the standard basis. It states that the reflection sends **e1** to **e2** and **e2** to **e1**.

Question 5. Consider the linear transformation $L : R^3 \to R^3$, rotating the plane x + y + z = 0 by an angle of $\pi/2$. Start with the basis $V = \{v1 = (\frac{1}{\sqrt{3}}, \frac{1}{\sqrt{3}}, \frac{1}{\sqrt{3}})$, $v2 = (\frac{-1}{\sqrt{2}}, \frac{1}{\sqrt{2}}, 0)$, $v3 = (\frac{-1}{\sqrt{6}}, \frac{-1}{\sqrt{6}}, \frac{2}{\sqrt{6}})\}$. The first vector is along the axis of rotation and L will send it to itself. The second and third vectors are unit vectors in the plane of rotation and are perpendicular to each other. Thus,

$$L(v1) = v1, \quad L(v2) = v3, \quad L(v3) = -v2.$$

(a) Obtain the matrix $A = [L]_V^V$ and enter it into Maple.

(b) Find the transition matrix T_S^V where $S = \{e_1, e_2, e_3\}$ is the standard basis. Use Maple to find $T_V^S = (T_S^V)^{-1}$.

(c) Use the formula $[L]_S^S = T_S^V[L]_V^V T_V^S$ to find $B = [L]_S^S$.

(d) Use your answer from (c) to find L((2,1,3)). Note that L is just multiplication by $[L]_S^S$.

Example 5. Consider differentiation as a linear transformation $D : P^2 \to P^1$. To find, for the bases $S_2 = \{1, x, x^2\}$ in P^2, and $S_1 = \{1, x\}$ in P^1, the matrix associated with D, we compute

$$D(1) = 0, \; D(x) = 1, \; D(x^2) = 2x.$$

To get the matrix $A = D_{S_1}^{S_2}$ we write $0 = 0 \cdot 1 + 0 \cdot x$, $1 = 1 \cdot 1 + 0 \cdot x$, $2x = 0 \cdot 1 + 2 \cdot x$. Thus

> A := transpose(matrix(3,2, [0,0,1,0,0,2]));

Now consider the process of finding antiderivatives whose value at $x = 0$ is 0. Thus,

$$I_0(p) = \int_0^x p(t)dt.$$ We regard this as a linear transformation $I_0 : P^1 \to P^2$. Then we find

$I_0(1) = x$, $I_0(x) = \frac{x^2}{2}$. We find the matrix $B = [I_0]_{P^2}^{P^1}$ by writing $x = 0 \cdot 1 + 1 \cdot x + 0 \cdot x^2$, $\frac{x^2}{2}$

$= 0 \cdot 1 + 0 \cdot x + (\frac{1}{2}) \cdot x^2$.

Question 6. Enter the matrix B. Then consider the composition $I_0 D$. By computing its values on the basis $\{1, x, x^2\}$ and expressing the answer in terms of this basis, give the matrix C representing the composition with respect to S_2. Compare your answer with the product BA. Find the matrix E which represents the composition DI_0 in the basis S_1. Compare your answer with the product AB and justify what you have found.

Given a basis in a vector space, we can ascribe coordinates (with respect to this basis) to each vector. These tuples of coordinates make the original vector space, for some integer n, isomorphic to R^n. For example, the vector space M(2,2) has a standard basis

$$S = \left\{ E(1,1) = \begin{pmatrix} 1 & 0 \\ 0 & 0 \end{pmatrix}, E(1,2) = \begin{pmatrix} 0 & 1 \\ 0 & 0 \end{pmatrix}, E(2,1) = \begin{pmatrix} 0 & 0 \\ 1 & 0 \end{pmatrix}, E(2,2) = \begin{pmatrix} 0 & 0 \\ 0 & 1 \end{pmatrix} \right\}.$$

Given a 2 by 2 matrix, we can then find its coefficients with respect to this basis. For example, the matrix $A = \begin{pmatrix} 2 & 1 \\ -3 & -1 \end{pmatrix}$ has coefficients

$[A]_S = (2,1,-3,-1)$ with respect to the basis S. Now, consider the basis W $= \left\{ \begin{pmatrix} 1 & 0 \\ 0 & 1 \end{pmatrix}, \right.$

$\left. \begin{pmatrix} 1 & 0 \\ 0 & -1 \end{pmatrix} \begin{pmatrix} 0 & 1 \\ 1 & 0 \end{pmatrix} \begin{pmatrix} 0 & 1 \\ -1 & 0 \end{pmatrix} \right\}$. The transition matrix T_S^W is found by writing as columns the

4-tuples that express the matrices in W in terms of the basis S. Consequently,

$$T_S^W = \begin{pmatrix} 1 & 1 & 0 & 0 \\ 0 & 0 & 1 & 1 \\ 0 & 0 & 1 & -1 \\ 1 & -1 & 0 & 0 \end{pmatrix}.$$

The matrix T_W^S is just $(T_S^W)^{-1}$. With respect to the basis W, the coefficients of the matrix A are obtained as $[A]_w = T_W^S [A]_s$.

```
> As := vector([2,1,-3,-1]);
> TWs := matrix(4,4, [1,1,0,0,0,0,1,1,0,0,1,-1,1,-1,0,0]);
> TSw := inverse(TWs);
> Aw := evalm(TSw * As);
```

The answer should be (1/2, 3/2, -1, 2). We can check this by calculating the linear combination of vectors in the basis W represented by these coordinates.

```
> w1 := matrix([[1,0],[0,1]]);
> w2 := matrix([[1,0],[0,-1]]);
> w3 := matrix([[0,1],[1,0]]);
> w4 := matrix([[0,1],[-1,0]]);
> evalm(1/2*w1 + 3/2*w2 - w3 + 2*w4);
```

Now, consider the linear transformation

$$L : M(2,2) \rightarrow M(2,2), L(B) = \begin{pmatrix} 2 & -1 \\ 1 & 2 \end{pmatrix} B.$$

We first find the matrix representing L in the basis S. We renumber the basis elements in S as E_1, E_2, E_3, E_4, respectively, and compute L on each basis matrix. We find

$$L(E_1) = 2E_1 + E_3, L(E_2) = 2E_2 + E_4, L(E_3) = -E_1 + 2E_3, L(E_4) = -E_2 + E_4.$$

```
> L := matrix([[2,-1], [1,2]]);
> E1 := matrix([[1,0], [0,0]]);
> E2 := matrix([[0,1], [0,0]]);
> E3 := matrix([[0,0], [1,0]]);
> E4 := matrix([[0,0], [0,1]]);
> evalm(L &* E1);
> evalm(L &* E2);
> evalm(L &* E3);
> evalm(L &* E4);
```

Thus, the matrix $[L]_s^s = \begin{pmatrix} 2 & 0 & -1 & 0 \\ 0 & 2 & 0 & -1 \\ 1 & 0 & 2 & 0 \\ 0 & 1 & 0 & 2 \end{pmatrix}$. To find the matrix $[L]_w^w$ we use the formula $[L]_w^w$

$= T_w^s [L]_s^s T_s^w$.

```
> Lss := matrix(4,4, [2,0,-1,0,0,2,0,-1,1,0,2,0,0,1,0,2]);
> Lww := evalm(TSw &* Lss &* TWs);
```

If we wanted to find the coefficients of L(A) with respect to one of the bases, we could make use of the following formulas:

$$[L(A)]_s = [L]_s^s [A]_s, \ [L(A)]_w = [L]_w^w [A]_w, \ [L(A)]_w = T_w^s [L(A)]_s.$$

```
> LAs := evalm(Lss &* As);
> LAw := evalm(Lww &* Aw);
> LAw := evalm(TSw &* LAs);
```

Question 7. Consider the space $V = \text{Sym}(3,3)$ of symmetric 3x3 matrices. This has a standard basis S consisting of the following matrices:

$$S(1,1) = \begin{pmatrix} 1 & 0 & 0 \\ 0 & 0 & 0 \\ 0 & 0 & 0 \end{pmatrix}, \ S(1,2) = \begin{pmatrix} 0 & 1 & 0 \\ 1 & 0 & 0 \\ 0 & 0 & 0 \end{pmatrix}, \ S(1,3) = \begin{pmatrix} 0 & 0 & 1 \\ 0 & 0 & 0 \\ 1 & 0 & 0 \end{pmatrix},$$

$$S(2,2) = \begin{pmatrix} 0 & 0 & 0 \\ 0 & 1 & 0 \\ 0 & 0 & 0 \end{pmatrix}, \ S(2,3) = \begin{pmatrix} 0 & 0 & 0 \\ 0 & 0 & 1 \\ 0 & 1 & 0 \end{pmatrix}, \ S(3,3) = \begin{pmatrix} 0 & 0 & 0 \\ 0 & 0 & 0 \\ 0 & 0 & 1 \end{pmatrix}.$$

(a) Give, with respect to the basis S, the coordinates $[A]_s$ of the symmetric matrix A =

$$\begin{pmatrix} 1 & 3 & -2 \\ 3 & 1 & -4 \\ -2 & -4 & 2 \end{pmatrix}$$ regarded as a vector in V. Denote your answer as AS and enter it into Maple.

(b) Consider the linear transformation $L : M(3,3) \rightarrow V$ given by $L(A) = A + A^t$. For a basis in $M(3,3)$, take $E = \{E(1,1), E(1,2), E(1,3), E(2,1), E(2,2), E(2,3), E(3,1), E(3,2), E(3,3)\}$, where $E(i,j)$ has a 1 in the ij-position and 0 elsewhere. Find the matrix $[L]_S^E$. Denote your answer by LES and enter it into Maple.

(c) Enter the matrix $B = \begin{pmatrix} 1 & 2 & 3 \\ 4 & 5 & 6 \\ 7 & 8 & 9 \end{pmatrix}$ into Maple. Give the vector of coefficients $[B]_E$ and

enter it into Maple as BE. Find $[L(B)]_S$ by first computing $L(B)$ and then writing your answer in terms of the basis S. Denote your answer by LBS and enter it into Maple. Then find it using Maple via the formula $[L(B)]_S = [L]_S^E [B]_E$.

(d) Another basis for V is W, given by

$$\begin{pmatrix} 1 & 0 & 0 \\ 0 & 0 & 0 \\ 0 & 0 & 0 \end{pmatrix}, \begin{pmatrix} 1 & 1 & 0 \\ 1 & 0 & 0 \\ 0 & 0 & 0 \end{pmatrix}, \begin{pmatrix} 1 & 1 & 0 \\ 1 & 1 & 0 \\ 0 & 0 & 0 \end{pmatrix}, \begin{pmatrix} 1 & 1 & 1 \\ 1 & 1 & 0 \\ 1 & 0 & 0 \end{pmatrix}, \begin{pmatrix} 1 & 1 & 1 \\ 1 & 1 & 1 \\ 1 & 1 & 0 \end{pmatrix}, \begin{pmatrix} 1 & 1 & 1 \\ 1 & 1 & 1 \\ 1 & 1 & 1 \end{pmatrix}.$$

Find the matrix T_s^W and use Maple to find the matrix T_w^s.

(e) Find $[L(B)]_w$.

LAB 7

Orthogonality and the Gram-Schmidt Algorithm

Contents:

Maple Commands:

dotprod, evalm, GramSchmidt, innerprod, int, map, matrix, normalize, sqrt, transpose, vector, with(linalg), with(orthopoly)

Topics:

Dot product, Gram-Schmidt algorithm, inner products, orthogonal basis, orthogonal complement, orthonormal basis, QR decomposition

We will first be using the dot product in R^n. We assume, as usual, that vectors are column vectors. In this case, the dot product is given by the formula

$$\langle \mathbf{v}, \mathbf{w} \rangle = \mathbf{v}^t \mathbf{w}.$$

For example, if we have the vectors $\mathbf{v} = (1,3,2,-1)$, $\mathbf{w} = (1,7,-2,3)$, then we compute the dot product in Maple in either of the following ways.

```
> with(linalg):
> v := vector([1,3,2,-1]);
> w := vector([1,7,-2,3]);
> dotprod(v, w);
> evalm(transpose(v) * w);
```

The Gram-Schmidt algorithm uses projection, implemented by the dot product, to convert an arbitrary basis into an orthonormal basis.

Example 1. Given the R^3 basis vectors $\mathbf{v1} = (1,1,0)$, $\mathbf{v2} = (1,3,2)$, $\mathbf{v3} = (1,3,-1)$, we demonstrate the Gram-Schmidt algorithm for orthonormalizing this basis. The orthonormal basis vectors will be $\mathbf{w1}$, $\mathbf{w2}$, and $\mathbf{w3}$.

```
> v1 := vector([1,1,0]);
> v2 := vector([1,3,2]);
> v3 := vector([1,3,-1]);
```

Define **w1** to be **v1** and compute **w2** and **w3** as follows.

```
> w1 := v1;
> r1 := dotprod(w1, w1);
> w2 := evalm(v2 - w1*dotprod(v2, w1)/r1);
> r2 := dotprod(w2, w2);
> w3 := evalm(v3 - w1*dotprod(v3,w1)/r1 - w2*dotprod(v3,w2)/r2);
> r3 := dotprod(w3, w3);
```

The vectors **w1, w2**, and **w3** are now mutually orthogonal. We next normalize them, noting that we have already computed the square of the norm for each vector.

```
> q1 := evalm(w1/sqrt(r1));
> q2 := evalm(w2/sqrt(r2));
> q3 := evalm(w3/sqrt(r3));
```

Maple's *linalg* package contains the command **GramSchmidt** for orthogonalizing a list of vectors.

```
> gs := GramSchmidt([v1, v2, v3]);
```

The output of this command is assigned to the name "gs" and is a list of what we have computed above as the vectors **w1, w2**, and **w3**. Since these vectors are not unit vectors, we next apply the **normalize** command from the *linalg* package.

```
> GS := map(normalize, gs);
```

If Maple were being used in a "production" mode instead of a didactic mode, then the orthonormalization process could be accomplished by the more succinct syntax

```
> map(normalize, GramSchmidt([v.(1..3)]));
```

The **map** command causes the **normalize** command to be applied to each vector in the list returned by the **Gram-Schmidt** command. The notation v.(1..3) represents the sequence v1, v2, v3. Clearly, this notation is not limited to just sequences of length 3.

For instructional purposes, students might find it useful to implement the Gram-Schmidt process stepwise. In this case, the simplest way to do this would be to copy and paste the instructions entered above, and re-execute them in the Maple Worksheet. Of course, a special-purpose function could be written, but it would only serve until familiarity with the Gram-Schmidt process allowed the user to move on to more direct computations.

Question 1. Orthonormalize the vectors **v1, v2, v3** of Example 1 by using the **GramSchmidt** and **normalize** commands.

Question 2. For the vectors **v1** = (1,2,-1,), **v2** = (3,2,1,-1), **v3** = (-2,1,3,1) in R^4, copy and paste the Maple instructions used above for orthonormalization, and re-execute them for the vectors given here. Then, use the **GramSchmidt** and **normalize** commands to do the orthonormalization.

Example 2. We now orthonormalize a set of vectors in P^3, a process that requires the use of integration over the interval [-1,1] as the inner product. If we start with the four vectors 1, x, x^2, x^3, we can implement the steps of the Gram-Schmidt process as follows. Note that the analog of <**v**, **w**> is $\int_{-1}^{1} v \, w \, dx$.

```
> v1 := 1;
> v2 := x;
> v3 := x^2;
> v4 := x^3;
> w1 := v1;
> r1 := int(w1*w1, x = -1..1);
> w2 := v2 - w1*int(v2*w1, x = -1..1)/r1;
> r2 := int(w2^2, x = -1..1);
> w3 := v3 - w1*int(v3*w1, x = -1..1)/r1 - w2*int(v3*w2, x = -1..1)/r2;
> r3 := int(w3^2, x = -1..1);
> w4 := v4 - w1*int(v4*w1, x = -1..1)/r1
        - w2*int(v4*w2, x = -1..1)/r2 - w3*int(v4*w3, x = -1..1)/r3;
```

Caution: On all platforms, Maple automatically wraps long lines. Don't hit the **Enter** key before the complete line is typed. If you insist on breaking the line yourself, the instructions are platform specific. On the Macintosh, there is an **Enter** key and a **Return** key. The **Return** key is a line-feed and will break the line without sending it to Maple for execution. The **Enter** key sends instructions to Maple to begin execution. Under any Windowing system (PC and UNIX) there is only one key. The line-feed is sent by **Shift-Enter**. However, on these platforms a second prompt will be displayed at the beginning of the broken line. On the Macintosh, one can configure Maple so that only one prompt shows for a broken line.

The polynomials w1, w2, w3, and w4 are orthogonal with respect to integration as the inner product on the interval [-1,1]. They are not normalized, and hence, are monic—the coefficient of the highest power term in each polynomial is 1.

Question 3. Do you recognize the set of monic polynomials being generated? Look up the various known sets of orthogonal polynomials such as Legendre polynomials, Hermite polynomials, Laguerre polynomials, Chebyshev polynomials, and so on. Maple has a collection of known orthogonal polynomials in its *orthopoly* package that you can access and then learn about by the commands

```
> with(orthopoly);
> ?orthopoly
```

Access some of these polynomials and see if you can tell if what was generated in Example 2 is one of these known sets of orthogonal polynomials.

Question 4. Find an orthogonal basis for the subspace of P^3 with basis $\{1 - x, x + x^2 + x^3, x^2 - x^3\}$.

Question 5. If the matrix $A = \begin{pmatrix} 2 & 0 & \frac{2}{3} & 0 \\ 0 & \frac{2}{3} & 0 & \frac{2}{5} \\ \frac{2}{3} & 0 & \frac{2}{5} & 0 \\ 0 & \frac{2}{5} & 0 & \frac{2}{7} \end{pmatrix}$ is defined, then the inner product

determined by $<v, w> = v^t A w$ can be used for the orthogonalization of Question 4. Represent the four basis polynomials $1, x, x^2, x^3$ as the vectors e_1, e_2, e_3, and e_4 of the standard basis in R^4. Show that $<x^i, x^j> = \int_{-1}^{1} x^{i+j} dx$ yields the same values as the inner product $e_i^t A e_j$. Hint: Enter the matrix A and the basis vectors via

```
> A := matrix(4,4, [2,0,2/3,0,0,2/3,0,2/5,2/3,0,2/5,0,0,2/5,0,2/7]);
> id := diag(1$4);
> for k from 1 to 4 do e.k := col(id, k); od;
```

Next, realize the inner product in Maple with the command **innerprod** from the *linalg* package. Typical usage would be

```
> innerprod(e1, A, e2);
```

Then implement the Gram-Schmidt orthogonalization process in terms of the vectors e_1, e_2, e_3, and e_4 and the inner product defined by A. This inner product replaces the dot product of Example 1.

The next question involves finding the orthogonal complement of the subspace spanned by the three vectors $\{(2,-1,0,2,1), (1,-3,0,0,1), (2,1,2,1,1)\}$ in R^5.

Question 6. (a) Form a matrix A with these three vectors as the rows. The row space of A is the subspace in question. Find a basis for it.
(b) The orthogonal complement is then the null space of A. Find a basis for N(A) by using each of the following options: linsolve(A, **0**), nullspace(A), rref(A).

(c) Form $B = A^t$. Now the subspace, which was the row space of A, becomes the column space of B. Its orthogonal complement is the null space of B^t. Check this by using **linsolve, nullspace**, and **rref**.

LAB 8

The Least Squares Problem

Contents:

Maple Commands:
augment, col, display, dotprod, equal, evalf, evalm, for/do/od, GramSchmidt, inverse, leastsqrs, leastsquare, linsolve, map, matrix, norm, normalize, op, plot, print, rand, randvector, rank, rref, _seed, seq, solve, subs, sum, transpose, vector, with(fit), with(linalg), with(plots), with(stats), ->, $

Topics:
Data fitting, least squares problem, normal equation, projections, regression line

At the heart of a solution to a least squares problem for an overdetermined system A**x** = **b** is the notion of a projection onto a subspace. It is therefore useful to explore the idea of a projection of a vector **x** onto the subspace spanned by the vectors $\mathbf{v_1}, \ldots, \mathbf{v_k}$. It should be noted that the failure of the spanning set to be linearly independent poses an additional complication.

We create, in Maple, an example that illustrates projection onto a subspace. This projection can be accomplished in two ways. First, by using techniques similar to those in the Gram-Schmidt orthogonalization process, we simply subtract from the given vector **x** any component along each vector $\mathbf{v_j}$, j = 1,..., k. The remaining component of **x** is then orthogonal to the subspace spanned by the **v**'s. The component of **x** in the subspace spanned by the **v**'s is the sum of all the components of **x** that were subtracted from **x**; hence, $\mathbf{x} = \mathbf{x_v} + \mathbf{x_\perp}$, where $\mathbf{x_v}$ is the component in the span of the **v**'s and $\mathbf{x_\perp}$ is the component orthogonal to this subspace. Second, we can construct a projection matrix P for which $\mathbf{x_v} = P\mathbf{x}$. We begin the example by demonstrating how to use Maple to generate examples at random.

```
> with(linalg):
> _seed := 1995;
> f := rand(-3..3):
```

We have seeded the random number generator with the starting integer 1995. Any positive integer can be used to condition the fixed sequence of pseudorandom numbers

Maple then generates. The **rand** function will ascribe to f random integers in the interval [-3,3]. We next randomly create **x**, **v1**, **v2**, **v3**, and **v4**, five vectors in R^5. It is unlikely that the vectors **v1**,…, **v4** will be linearly dependent, but we illustrate how Maple can be used to test for independence.

```
> for k from 1 to 4 do v.k := randvector(5, entries = f); od;
> x := randvector(5, entries = f);
> A := augment(v.(1..4));
> rank(A);
```

Projection is simplest when the spanning set is orthogonal. The key idea is the representation of x_v as a linear combination of the **v**'s. Hence, $x = x_{\perp} + x_v = \sum_1^4 c_k v_k$ and $\langle x, v_j \rangle = \langle x_v, v_j \rangle$ since $\langle x_{\perp}, v_j \rangle = 0$; moreover, $\langle x_v, v_j \rangle$ reduces to $c_j \langle v_j, v_j \rangle$ if the **v**'s are orthogonal. This gives $c_j = \dfrac{\langle x, v_j \rangle}{\langle v_j, v_j \rangle}$ from which follows $x_v =$

$\sum_1^4 \dfrac{\langle x, v_j \rangle}{\langle v_j, v_j \rangle} v_j$. We begin, then, by applying the Gram-Schmidt algorithm to the **v**'s.

```
> q := GramSchmidt([v.(1..4)]);
```

The orthogonalized versions of the **v**'s now reside in the list q from which the jth member can be recalled by the notation q[j]. The quotes on the argument to **sum** keep Maple from evaluating the dot products before an index value has been provided by the summation syntax, and the quotes on the summation index are necessary since k exits the earlier for-loop assigned the value 5 (much as in FORTRAN).

```
> xv := evalm(sum('dotprod(x, q[k])/dotprod(q[k], q[k])*q[k]', 'k' = 1..4));
> x_perp := evalm(x - xv);
```

We can test that $x_{\perp}$ (given in Maple by x_perp) is indeed orthogonal to each of the **v**'s by the computation

```
> for k from 1 to 4 do dotprod(x_perp, v.k); od;
```

We next construct the projection matrix P from the orthonormalized versions of the **v**'s. This requires that we map the **normalize** command onto the elements of the list q which already contains the orthogonalized versions of the **v**'s. These orthonormalized vectors can then be made the columns of a matrix Q by the **augment** command. Since this command takes a sequence, not a list, as argument, we use the **op** (operand) command to extract, as a sequence, the members of the list.

> Q := augment(op(map(normalize, q)));

The matrix P is then QQ^t and we test to see if Px is indeed x_v.

> P := evalm(Q &* transpose(Q));
> evalm(P*x);

Question 1. Project the vector $x = (3,-4,2,0,7)$ onto the subspace spanned by $v_1 = (1,-1,1,2,3)$, $v_2 = (-1,0,2,1,1)$, $v_3 = (0,1,-1,2,2)$. Find the component of x that is orthogonal to this subspace.

Question 2. Consider the Gram-Schmidt process applied to $v_1 = (1,1,0,1)$, $v_2 = (-1,1, 1,0)$, $v_3 = 2v_1 + 3v_2$, $v_4 = (0,1,0,2)$, a set of vectors that is not linearly independent. First, simply apply Maple's **GramSchmidt** command. Note that Maple outputs just three vectors, having detected that the set is not independent. Explain this loss of a vector in terms of the subtraction of components that takes place in the orthogonalization process.

Question 3. Orthogonalize the four vectors in Question 2 by implementing the steps of the Gram-Schmidt process. Show that the linear dependence causes w_3 to be 0, with $<w_3, w_3> = 0$. Hence, the algorithm cannot compute w_4 without "operator intervention." What can you conclude about Maple's **GramSchmidt** command?

We have constructed the projection matrix P as QQ^t, with the columns of Q being the orthonormalized versions of the basis for the subspace into which the projection is to be made. The projection matrix can also be constructed directly from the basis vectors without orthonormalizing if the basis vectors are made to be the columns of a matrix A and we compute $P = A(A^tA)^{-1}A^t$. For the vectors in Example 1 we verify that this yields the same projection matrix as QQ^t. Since the matrix A already exists, we need only compute

> P1 := evalm(A &* inverse(transpose(A) &* A) &* transpose(A));
> equal(P, P1);

Question 4. Demonstrate that if A is identified with the matrix Q, and we form $Q(Q^tQ)^{-1}Q^t$, the matrix being inverted, namely, Q^tQ, reduces to the 4x4 identity matrix. Hence, $Q(Q^tQ)^{-1}Q^t$ reduces to QQ^t, just as before.

What happens if we try to compute the projection matrix as $P = A(A^tA)^{-1}A^t$ when the columns of A are linearly dependent? Well, A is not of full rank because its columns are dependent. Therefore, there are vectors in the null space of A, that is, $Ax = 0$ for some nonzero x. Thus, $(A^tA)x = A^t(Ax) = A^t0 = 0$, so there are also nonzero vectors in the null space of A^tA. So the required inverse does not exist, and the projection matrix P cannot be calculated by this formula. The only way to construct such a projector P is by orthonormalizing the columns of A to form Q. But this can only be done by a process that recognizes the linear dependence of the columns of A and eliminates the "excess" vectors, as Maple's **GramSchmidt** command would.

We now have enough machinery in place to tackle the least squares solution of the inconsistent system $Ax = b$. We project **b** onto the column space of A and call this projection b_A. Since the vector b_A now lies completely in the column space of A, the system $Ax = b_A$ has a solution. That is the solution we seek, and when we find it we will call it the "least squares solution."

Example 1. Define the following matrix A and vector **b** for the inconsistent system $Ax = b$.

```
> A := matrix(5,3, [-2,1,-1,1,3,2,-1,-2,-1,0,1,1,1,2,1]);
> b := vector([1,2,3,4,5]);
> A1 := augment(A, b);
> rank(A);
> rank(A1);
> rref(A1);
```

We discover that A has maximal rank of 3 and the augmented matrix A1 has rank 4, indicating that **b** is indeed not in the column space of A. The equations are inconsistent, as the reduced normal form rref(A1) shows. We indeed need a least squares solution which we implement through the projection matrix $P = A(A^tA)^{-1}A^t$. P can be found by this construction since the columns of A are independent.

```
> P := evalm(A &* inverse(transpose(A) &* A) &* transpose(A));
> b_A := evalm(P*b);
> x_ls := linsolve(A, b_A);
```

In this case, we obtain the unique least squares solution x_ls. In our next example, the columns of A will be linearly dependent, and we will discover some of the difficulties thereby generated.

Example 2. Change the (1,1)-element in the matrix A to 2, call the new matrix B, and retain the same vector **b** to form the inconsistent system $Bx = b$. Since it is not obvious how to transform the existing matrix A to the desired matrix B, we will illustrate that technique.

```
> B := A;
> B[1,1] := 2;
> print(B);
```

The alternative is to enter B in its entirety.

> B := matrix(5,3, [2,1,-1,1,3,2,-1,-2,-1,0,1,1,1,2,1]);

In either event, the Maple steps remaining are as follows.

> b := vector([1,2,3,4,5]);
> B1 := augment(B,b);
> rank(B);
> rank(B1);
> rref(B1);

We should discover that the rank of B is 2 but the rank of the augmented matrix B1 is 3. This says there are two independent columns in B, but 3 in B1, so the column **b** in B1 is linearly independent of the columns of B. The equations are inconsistent, as can also be seen from the reduced normal form produced by rref(B). We really need to obtain a least squares solution. And since the columns of B are not independent, the projection matrix P has to be formed by invoking the Gram-Schmidt process. Since the column vectors spanning the subspace (the column space of B) are already in a matrix, we need to extract them first before we can hand them to Maple's **GramSchmidt** command, which itself returns a list. The **augment** command takes a sequence, so we again extract the orthonormalized vectors from the list q1 with the **op** command.

> q := [seq(col(B,k),k=1..3)];
> q1 := map(normalize, GramSchmidt(q));
> Q := augment(op(q1));
> P := evalm(Q &* transpose(Q));
> b_B := evalm(P*b);
> x_ls := linsolve(B, b_B);

This time, the least squares solution x_{ls}, given in Maple by x_ls, is not unique because the rank of B was not maximal. Additional computation is required to project x_{ls} onto the row space of B. The least squares solution always resides in the row space of the original matrix, here B. That is why we will project x_{ls} onto this row space. And that requires us to repeat, for B^t, the above Gram-Schmidt calculations.

> B1 := transpose(B);
> q := [seq(col(B1,k),k=1..3)];
> q1 := map(normalize, GramSchmidt(q));
> Q := augment(op(q1));
> P1 := evalm(Q &* transpose(Q));
> x_rs := evalm(P1*x_ls);

If x_{rs}, x_{ls}, and b_B are respectively represented in Maple by x_rs, x_ls, and b_B, we verify that x_{rs}, the projection of x_{ls} onto the row space of B, satisfies the equation $Bx_{rs} = b_B$.

> evalm(B*x_rs - b_B);

To test that $\mathbf{x_{rs}}$ is in the row space of B^t, do either of the following computations.

> rref(augment(B1, x_rs));
> rank(B1);
> rank(augment(B1, x_rs));

Since $\mathbf{x_{rs}}$ is the projection of $\mathbf{x}$ onto the row space of B, B takes $\mathbf{x_{rs}}$ to $\mathbf{b_B}$, the projection of $\mathbf{b}$ onto the column space of B. In fact, $B\mathbf{x_{rs}} = \mathbf{b_B}$. The vector $\mathbf{b} - \mathbf{b_B}$ is orthogonal to the column space of B, and its length is a measure of how close $\mathbf{b_B}$ is to $\mathbf{b}$. This length can be computed as

> norm(b_B, 2);

where the parameter 2 indicates Maple is to compute the 2-norm, the square root of the sum of the squares of the components, equivalent to the square root of the dot product of $\mathbf{b_B}$ with itself.

Note that Maple's *linalg* package has a **leastsqrs** command, the use of which in Example 1 produces the same unique solution found before. However, its use on the system in Example 2 gives the nonunique solution $\mathbf{x_{ls}}$ which still needs to be projected onto the row space of B.

> leastsqrs(B, b);

Question 5. Use the appropriate projection matrix P to obtain the least squares solution to the system $A\mathbf{x} = \mathbf{b}$, where

$$A = \begin{pmatrix} 1 & -1 & -1 \\ 0 & 1 & 3 \\ 2 & 1 & 7 \\ 1 & 2 & 8 \end{pmatrix}, \mathbf{b} = \begin{pmatrix} 1 \\ 1 \\ 1 \\ 1 \end{pmatrix}.$$

Consider next an example of obtaining the least squares solution by the use of the normal equations.

Example 3. If $A = \begin{pmatrix} 1 & 2 \\ -1 & 3 \\ 0 & 1 \end{pmatrix}$ and $\mathbf{b} = \begin{pmatrix} 1 \\ 1 \\ 1 \end{pmatrix}$, obtain the least squares solution of $A\mathbf{x} = \mathbf{b}$ via the normal equations $A^tA\mathbf{x} = A^t\mathbf{b}$.

```
> A := matrix(3,2, [1,2,-1,3,0,1]);
> b := vector([1,1,1]);
> At : transpose(A);
> B := evalm(At &* A);
> c := evalm(At * b);
> x_ls := linsolve(B, c);
```

Question 6. Obtain the solution of Example 3 by use of the **leastsqrs** command. Show that x_{ls} is indeed in the row space of A, and explain why this is so.

Let us obtain the least squares solution in Example 3 by constructing the appropriate projection matrix P.

```
> P := evalm(A &* inverse(transpose(A) &* A) &* transpose(A));
> b_A := evalm(P*b);
> linsolve(A, b_A);
```

We were able to obtain P without use of the Gram-Schmidt orthogonalization process because A has maximal rank of 2.

Question 7. Obtain the least squares solution in Example 1 by using the normal equations.

Question 8. Obtain the least squares solution in Example 2 by using the normal equations. Do the normal equations avoid the need to project back onto the row space of B?

Least squares calculations arise, for example, when fitting curves to data. If there are exactly the same number of data points as there are parameters in the curve, the problem is one of interpolation, and the fitting curve passes through each of the data points. When there are more points than parameters in the fitting curve, the system of equations specifying the parameters will be overdetermined and inconsistent. That is when a least squares solution is appropriate.

Example 4. Fit the given three points to a quadratic polynomial. Note that there are exactly as many parameters (a, b, c) as there are points, so there are exactly three equations in three unknowns. The system is determinate, and the parabola interpolates the three points.

```
> P1 := [1,-3];
> P2 := [3, 5];
> P3 := [-2, 4];
> Y := a + b*x + c*x^2;
> for k from 1 to 3 do e.k := subs(x = P.k[1], Y) = P.k[2]; od;
> q := solve({e1, e2, e3}, {a, b, c});
> Y1 := subs(q, Y);
```

Example 5. Fit a straight line to the above data points. This time, there are only two parameters, so there will be three equations in only two unknowns. The system of equations for these two unknown parameters is overdetermined. In fact, these equations are

```
> b := 'b';
> Y := a + b*x;
> for k from 1 to 3 do e.k := subs(x = P.k[1], Y) = P.k[2]; od;
```

The **leastsqrs** command in the *linalg* package will accept the least squares problem in a matrix/vector form or in the present form of equations. In this latter case, the output from the **leastsqrs** command is a set of equations defining the parameters.

```
> q := leastsqrs({e1, e2, e3}, {a, b});
> Y1 := subs(q, Y);
```

If we want to see a plot of the data points along with a graph of the regression line Y1, we can make use of the **display** command in the *plots* package for superimposing plots.

```
> with(plots):
> P := [P.(1..3)];
> f1 := plot(P, style = point, symbol = circle, color = red):
> f2 := plot(Y1, x = -3..5):
> display([f1, f2]);
```

The list of points in P is a list-of-lists. If, for some reason, the data came in such a format, it could be "flattened" to two separate lists of x- and y-coordinates as follows.

```
> X_data := map(z -> z[1], P);
> Y_data := map(z -> z[2], P);
```

The matrix form of the least squares problem can be generated from the lists X and Y as follows.

```
> U := [1$3];
> A := augment(U, X_data);
> Y := vector(Y_data);
> q := leastsqrs(A, Y);
```

This time, q is a vector whose two entries are the values of the parameters a and b in the regression line $y = a + bx$. U is a list of 1's, the coefficients of a in each of the equations e1, e2, and e3.

The least squares problem in matrix form also admits a solution by the normal equations.

```
> AtA := evalm(transpose(A) &* A);
> AtY := evalm(transpose(A) * Y);
> linsolve(AtA, AtY);
```

Maple also has the **leastsquare** command in the *fit* subpackage of its *stats* package. The output from this command will be the actual equation of the fitting curve, not just the coefficients. Hence, this command applies to fitting functions with the linear form $\sum_{k=0}^{n} a_k \varphi_k(x)$. The data needs to be in two lists, one list of x-coordinates and one of y-coordinates.

Example 6. Use the **leastsquare** command to fit $y = a \sin(x) + b \cos(x)$ to the data of Example 4.

```
> with(stats):
> with(fit):
> q := leastsquare[[x,y], y = a*sin(x) + b*cos(x),{a, b}]([X_data, Y_data]);
> evalf(q);
```

Question 9. For the data (x,y) =

$$(1,3), (2,4), (3,2), (4,3), (5,5), (6,5), (7,4), (8, 7/2), (9,3), (10,2)$$

(a) find the regression line;
(b) find the cubic $y = a_0 + a_1x + a_2x^2 + a_3x^3$ which best fits the data;
(c) find the polynomial of degree no more than 9 which interpolates these data points;
(d) plot the data points and these three curves on the same graph.
For parts (a) and (b), solve by formulating the least squares problem in matrix notation and using the **leastsqrs** command and the normal equations. Also, formulate the problem as a set of equations and use **leastsqrs** on these equations. Finally, use the **leastsquare** command.

LAB 9

Eigenvalue-Eigenvector Problems

Contents:

Maple Commands:
allvalues, augment, charpoly, convert/float, convert/rational, det, eigenvals, eigenvects, evalf, evalm, inverse, map, matrix, nullspace, op, randmatrix, _seed, solve, sort, subs, trace, with(linalg)

Topics:
Characteristic polynomial, eigenvalue-eigenvector problem, relation of eigenvalues to trace and det, similar matrices

This lab uses Maple to solve eigenvalue-eigenvector problems. There are two fundamental tools for this in the *linalg* package, the commands **eigenvals** and **eigenvects**. The **eigenvals** command returns a sequence of the eigenvalues, or their equivalent. The **eigenvects** command returns a sequence of lists, each list having three members: the eigenvalue, the algebraic multiplicity of that eigenvalue, and a set of the eigenvectors belonging to that eigenvector.

Each command operates in two modes, symbolic and numeric. If the entries of a matrix are all symbolic (and this includes the case of entries being rational numbers), then Maple attempts to compute the eigenvalues or eigenvectors in exact, symbolic form. If a single entry of the matrix is a floating point number, then the whole computation is performed numerically and a floating point result is returned. Hence, caution must be exercised when executing either of these commands on matrices that would have Maple attempting to solve exactly a characteristic polynomial of degree greater than four.

There is one other interesting feature of both of these commands. Maple uses the *RootOf* data structure for expressing algebraic numbers. The output of both **eigenvals** and **eigenvects** can be in terms of such *RootOf* structures. However, both commands take an optional parameter, radical, which forces the return to be in terms of radicals and not the *RootOf* structure. However, this parameter cannot accomplish what is unobtainable. A 10x10 matrix will most likely have its eigenvalues expressed by a RootOf structure, and urging Maple to express the eigenvalues in radicals would not necessarily be successful.

Example 1. Find the eigenvalues and eigenvectors of the matrix

$$A = \begin{pmatrix} 11 & -4 & -2 \\ 2 & 2 & -2 \\ 4 & -2 & 5 \end{pmatrix}.$$

```
> with(linalg):
> A := matrix(3,3, [11,-4,-2,2,2,-2,4,-2,5]);
> qe := eigenvals(A);
> q := eigenvects(A);
```

Notice that for this matrix, Maple was able to find the integer eigenvalues without resort to either floating point arithmetic or use of the *RootOf* data structure. In qe we have the sequence of three eigenvalues. In q we have a sequence of three lists. The first list contains the eigenvalue 6, the algebraic multiplicity 1, and a set containing the one eigenvector **v1** = (0,1,-2). Each of the other two lists in q has a similar structure. To access the eigenvalues and eigenvectors from q, we use the following "selection" syntax.

```
> x1 := q[1][1];
> x2 := q[2][1];
> x3 := q[3][1];
> v1 := q[1][3][1];
> v2 := q[2][3][1];
> v3 := q[3][3][1];
```

The eigenvalues x1, x2, x3 are the first members of each of the three lists in the sequence q. The eigenvectors **v1**, **v2**, **v3** are each in a set that is the third member of each of the three lists in q. Each eigenvector is the first member in the set that contains them.

Example 2. For the matrix A of Example 1, compute the eigenvalues and eigenvectors "by hand" with Maple. Find the transition matrix for the similarity transformation that diagonalizes A.

```
> cp := charpoly(A, x);
> cp1 := det(A - x);
> sort(cp1);
```

Notice that Maple's **charpoly** command yields the characteristic polynomial equivalent to det(A - xI). Maple allows this determinant to be entered without the identity matrix I, and then writes the terms of this polynomial in a random order. If the **sort** command is used to rearrange the terms in this polynomial, we obtain the negative of the polynomial cp. The lead coefficient in cp is 1, the second coefficient is -trace(A), and the last one is $(-1)^n \det(A)$.

```
> trace(A);
> det(A);
```

Of course, we can recover the eigenvalues of A by solving the characteristic polynomial for x.

```
> solve(cp = 0, x);
```

Each eigenvector of the eigenvalue x_k is a member of the null space of the matrix A - x_kI. Hence, with qe being the sequence of eigenvalues 6, 3, 9, we find the corresponding eigenvectors by

```
> v1 := op(nullspace(A - x1));
> v2 := op(nullspace(A - x2));
> v3 := op(nullspace(A - x3));
```

The **nullspace** command returns a basis for the null space as a set of vectors. The **op** (operand) command extracts the single eigenvector from the set braces supplied by **nullspace**.

The transition matrix for the change of basis in which A is diagonalized has as its column vectors the eigenvectors **v1, v2, v3**. We put these vectors into a matrix S and demonstrate the similarity transformation that takes A into a diagonal matrix.

```
> S := augment(v1, v2, v3);
> d := evalm(inverse(S) &* A &* S);
```

The matrix d is a diagonal matrix with the eigenvalues along the main diagonal, and in the order corresponding to their eigenvectors in S. (We again point out that in Maple, D is a differentiation operator and is hence a reserved symbol.)

The matrix A was very carefully chosen to have integer eigenvalues and eigenvectors. If we select a 3x3 matrix at random, we will most likely find the eigenvalues and eigenvectors much more complicated.

Example 3. Set the internal variable _seed to 1995 and obtain the eigenvalues and eigenvectors of the random matrix this generates.

```
> _seed := 1995;
> A := randmatrix(3,3);
> qe := eigenvals(A);
```

The eigenvalues are expressed as radicals, but there is a complex conjugate pair and the expressions are complicated. They make use of Maple's %-abbreviations whereby repeated expressions are assigned the placeholders %n, n = 1,2,..., to compress the output. While this feature can be controlled or curtailed by the user, the expressions will not get any simpler. Hence, we resort to a purely numeric solution. We can either float the existing expressions or recalculate completely in floating point arithmetic.

```
> evalf(qe);
> Af := subs(A[1,1] = 14., op(A));
> eigenvals(Af);
> eigenvects(Af);
```

The substitution of one entry of A, here the (1,1)-element, for its floating point equivalent, causes the complete calculation of both eigenvalues and eigenvectors to be done numerically, in floating point arithmetic.

An alternative syntax for rendering A as a floating point matrix has Maple converting all the entries in A to their decimal equivalents.

```
> map(convert, A, float);
```

No discussion of Maple's eigen-solvers is complete without an experience with output expressed as *RootOf* structures.

Example 4. Compute the eigenvalues and eigenvectors of

$$A = \begin{pmatrix} 2 & 0 & -1 \\ 2 & 1 & -2 \\ -3 & 1 & 1 \end{pmatrix}.$$

```
> A := matrix(3,3, [2,0,-1,2,1,-2,-3,1,1]);
> qe := eigenvals(A);
> q := eigenvects(A);
```

The eigenvalues have been computed in terms of radicals, but the eigenvectors have been given in terms of the *RootOf* notation. We can either convert the *RootOf* structure to radicals or recalculate, as radicals, the eigenvectors themselves. The second alternative is simplest.

```
> eigenvects(A, radical);
```

The optional parameter, radical, successfully forced the **eigenvects** command to return its results in terms of radicals. Extracting the equivalent answer from the *RootOf* structure in q is more subtle. First, you will have to inspect the sequence q to determine if Maple has ordered the members of the sequence consistently. Here, Maple has put the list with the *RootOf* structure last. It could as well have been first. In fact, re-executing the command could cause Maple to re-order the terms in the sequence q. That is why the user must intervene and inspect the Maple output to determine which list in the sequence q contains the *RootOf* structure.

```
> allvalues(q[2]);
```

This yields a sequence of four lists, each with an eigenvalue and eigenvector. Clearly, this is wrong. The difficulty resides in the appearance of two *RootOf*'s in the list q[2]. When each of these is converted to radicals, Maple creates all possible pairings and gets four. To force Maple to keep equivalent roots together, issue the **allvalues** command with the optional parameter 'd'. The quotes are necessary since the variable d has been assigned a value earlier. The use of this optional parameter causes the **allvalues** command to return just the distinct pairings arising from the different *RootOf* structures.

> allvalues(q[2], 'd');

Question 1. For each of the matrices given below,
(a) find the characteristic polynomial and its roots;
(b) find the eigenvectors by use of the **nullspace** command;
(c) find the eigenvalues and eigenvectors by use of eigenvals and eigenvects;
(d) find the transition matrix S for the similarity transformation that diagonalizes A, and diagonalize A with it.

$$(1) \begin{pmatrix} -5 & 2 & 1 \\ -4 & 2 & 0 \\ -7 & 2 & 3 \end{pmatrix}, \quad (2) \begin{pmatrix} 3 & 0 & 1 & 0 \\ 0 & -1 & 0 & 3 \\ 1 & 0 & 3 & 0 \\ 0 & 3 & 0 & -1 \end{pmatrix}, \quad (3) \begin{pmatrix} 1 & 1 \\ 0 & 1 \end{pmatrix}, \quad (4) \begin{pmatrix} 1 & -1 \\ 1 & 1 \end{pmatrix}$$

Question 2. Construct a matrix A which has eigenvalues 1, 2, 3, 4, and corresponding eigenvectors $v_1 = (1,0,1,0)$, $v_2 = (1,-1,1,-1)$, $v_3 = (0,2,0,1)$, $v_4 = (0,2,0,1)$. (The eigenvectors are the columns of the transition matrix S that diagonalizes A by a similarity transformation $S^{-1}AS = D$. The diagonal matrix D has the eigenvalues on its main diagonal, but be sure not to use the letter D as a variable in Maple.)

Question 3. Let A = $\begin{pmatrix} 1 & -0.45 & 0.9 & -0.45 \\ 1.6 & -0.8 & 2 & -0.7 \\ 0.8 & -0.55 & 1.3 & -0.25 \\ 0 & -0.2 & 0.4 & 0.3 \end{pmatrix}$. Compute $R = A^{1000}$. Because of

round-off error, this computation could be difficult in a strictly numerical environment. Examine the following alternative way to evaluate R. Use map(convert, A, rational) to convert the entries of A to rational numbers. Recompute R in exact arithmetic. Don't be surprised at the size of the rational numbers so generated! Since the **evalf** command takes, as a second argument, the number of digits to be used in the evaluation, it is possible to determine R to a very high degree of precision. Was the conversion to rational arithmetic necessary?

LAB 10

Complex Eigenvalues and Eigenvectors

Contents:

Maple Commands: *abs, alias, argument, col, conjugate, convert/rational, det, diag, Digits, dotprod, eigenvects, equal, evalc, evalf, evalm, expand, fnormal, for/do/od, htranspose, I, Im, inverse, map, matrix, norm, op, print, Re, seq, sin, sqrt, vector, with(linalg), ->, @*

Topics: Adjoint, complex arithmetic, complex conjugation, Hermitian matrices, length of complex vectors, polar representation, powers of matrices, skew Hermitian matrixes, unitary matrices

Maple treats complex numbers like any other numbers. However, there are several special commands that pertain to the manipulation of complex quantities. Note, moreover, that Maple uses I for $\sqrt{-1}$ as a default definition. This can be changed by the user via the **alias** command which sets up notational equivalences that Maple recognizes both on input and on output. Initially, the definition for I is established by an **alias**. Hence, to declare that some other symbol, say j, is to stand for $\sqrt{-1}$ it is necessary to "unalias" I and then **alias** j to $\sqrt{-1}$.

```
> alias(I = I);
> alias(j = sqrt(-1));
> sqrt(-5);
```

Maple's response to the last input is $j\sqrt{5}$. But we will restore I as the alias for $\sqrt{-1}$ and keep the default definition from henceforth. However, in deference to standard mathematical exposition, we will use i when writing mathematical, not Maple, notation.

```
> alias(j = j, I = sqrt(-1));
```

Example 1. Addition, subtraction, multiplication, and division of complex numbers are handled by the standard operators of arithmetic, namely, +, -, *, and /. The magnitude of a complex number is returned by the **abs** function, and the complex argument (the angle in polar representation) is returned by the **argument** function, and lies in the range (-π, π]. Consequently, the addition of such arguments is only well-defined modulo 2π.

```
> a := 4 + 3*I;
> b := 2 - I;
> c := a + b;
> d := a*b;
> e := 2 - 2*b;
> f := a/b;
> abs(a);
> abs(b);
> abs(d);
> abs(a)*abs(d);
> argument(a);
> argument(b);
> argument(a) + argument(b);
> argument(d);
> e := -2*I;
> e^2;
> argument(e^2);
> 2*argument(e);
> (argument(e^2) - 2*argument(e))/(2*Pi);
```

Complex vectors are entered as any other vector in Maple.

```
> with(linalg):
> vector([2 + I, 3 - 2*I]);
```

Since Maple can entertain symbolic quantities that we intend to be complex, we can certainly create expressions for which Maple will not automatically carry out the intended complex arithmetic.

```
> z1 := 2*x + 3*I*y;
> z2 := 5*x - 7*I*y;
> expand(sin(z1)*z2);
> evalc(sin(z1)*z2);
```

The **expand** command merely multiples the binomials represented by the product $z2$ $\sin(z1)$, whereas the **evalc** (evaluate complex) command groups the real and imaginary parts of the product. In addition, one extracts the real and imaginary parts of complex quantities with the **Re** and **Im** commands. However, these commands do not compute anything unless they are the argument of a call to **evalc**.

```
> qr := Re(z1*z2);
> qi := Im(z1*z2);
> evalc(qr);
```

> evalc(qi);

Question 1. Perform the following operations of complex arithmetic.
(a) (-4 -2*i*)(-2 + 3*i*)
(b) (7 - 3*i*) + 3(2 + *i*)
(c) (-7 + *i*)/(2 - 5*i*)

Question 2. (a) Show that when we multiply the complex numbers 4 - 2*i* and 2 + 3*i* above, the lengths multiply and the arguments add (up to a multiple of 2π).
(b) Show that if we divide -7 + *i* by 2 - 5*i*, the length of the quotient is the quotient of the lengths and the argument of the quotient is the difference of the arguments (up to a multiple of 2π).

  Maple's **dotprod** command correctly handles complex vectors, but note that it conjugates the *second* argument! Recall that the conjugate of the complex number $z = x + i y$ is $x - i y$ and is typically denoted by $\bar{z}$.

> z := x + I*y;
> q := vector([1,1]);
> p1 := dotprod(q, z*q);
> p2 := dotprod(z*q, q);

The scalar p1 shows that the second argument, the term z*q, was conjugated. The scalar p2 shows no evidence that z was conjugated. Of course, the vector q was deliberately chosen as real to have the effects of this experiment attached to z. Many texts define the complex inner product with conjugation being applied to the *first* argument. Thus, if a text such as Lawson's <u>Linear Algebra</u> takes <**x,y**> as $\mathbf{x}^{*}\mathbf{y}$, then we would say that the *first* entry is conjugated. If a text or program defines <**x,y**> as the dot product between **x** and $\bar{\mathbf{y}}$, then we would say that the *second* argument is conjugated. It is this latter that we find Maple doing. Care must be taken to determine the convention of the particular text or program being used.
  Maple conjugates complex numbers with its **conjugate** function.

> conjugate(2 + I);
> conjugate(z);
> evalc(conjugate(z));

In the first case, the conjugation is immediate. In the second, it takes an **evalc** to "activate" the complex operation.
  To apply conjugation to the components of a complex vector, the **map** command must be used.

> q1 := vector([1 + I, 2 + 3*I]);
> map(conjugate, q1);

Should it be required after a conjugation, an **evalc** can be mapped onto the components of a vector as a composition of functions. Such compositions are implemented in Maple with the @ symbol.

```
> ?@
> q2 := vector([z, z]);
> map(evalc@conjugate, q2);
```

The difference in the two cases is the symbolic nature of the vector q2, as opposed to the numeric nature of the vector q1.

Maple distinguishes between the transpose and the conjugate-transpose (hermitian transpose) of a matrix or vector. It uses two separate commands, **transpose** and **htranspose**. When **htranspose** is applied to a complex vector, the conjugation is performed, but the transpose operation is characterized by the same screen display as we saw for the transpose of a real vector: the transposition takes place but is not displayed. It is merely indicated by the use of the word *transpose*, just as is the case with the **transpose** command.

```
> htranspose(q2);
```

The norm (or length) of a complex vector can be computed with the **norm** command, provided the 2-norm is indicated. The return will typically contain the absolute value function, shown on the screen as $|\cdots|$. Maple will convert this absolute value to its equivalent with a call to the **evalc** function.

```
> p := norm(q2, 2);
> evalc(p);
```

Question 3. Division of complex numbers can be expressed in terms of conjugation and magnitude. Verify in Maple that for

(a) the complex number a = 4 + 3i, $(1/a) = \dfrac{\bar{a}}{\|a\|^2}$;

(b) the numbers a and b = 2 - i, $a/b = \dfrac{\overline{ab}}{\|b\|^2}$.

Example 2. A matrix is Hermitian if it is equal to its conjugate transpose (adjoint), is skew-Hermitian if it is the negative of its conjugate transpose, and is unitary if its inverse is its conjugate transpose. Below, the matrix A is Hermitian, B is skew-Hermitian, C is unitary, and F is none of these.

```
> A := matrix(3,3, [2,-3+2*I,2-5*I,-3-2*I,1,4-3*I,2+5*I,4+3*I,3]);
> B := matrix(3,3, [2*I,-3+2*I,2-5*I,3+2*I,I,-4+3*I,-2-5*I,4+3*I,3*I]);
> C := matrix(2,2, [(1+I)/2, (1-I)/2, (1-I)/2, (1+I)/2]);
> F := matrix(2,2, [(1+I)/2, (1+I)/2, (1-I)/2, (1-I)/2]);
```

```
> equal(A, htranspose(A));
> equal(-B, htranspose(B));
> equal(inverse(C), htranspose(C));
> equal(F, htranspose(F));
> equal(-F, htranspose(F));
> equal(inverse(F), htranspose(F));
> det(F);
> htranspose(F);
> evalm(htranspose(F) - F);
> evalm(htranspose(F) + F);
> evalm(htranspose(F) &* F);
```

Question 4. Determine which of the following matrices are Hermitian, skew-Hermitian, or unitary.

$$A = \begin{pmatrix} 4i & 3+i \\ 3-i & 5i \end{pmatrix}, B = \begin{pmatrix} 4i & 3+i \\ -3+i & 5i \end{pmatrix}, C = \begin{pmatrix} 0 & 2+3i & 2-3i \\ 2-3i & 0 & 2+3i \\ 2+3i & 2-3i & 0 \end{pmatrix},$$

$$F = \begin{pmatrix} 0 & 2+3i & 2-3i \\ -2-3i & 0 & -2+3i \\ 2+3i & -2-3i & 0 \end{pmatrix}, G = \begin{pmatrix} \dfrac{1+2i}{3} & \dfrac{1-2i}{3} \\ \dfrac{\sqrt{2}}{3}(1-i) & \dfrac{\sqrt{2}}{3}(1+i) \end{pmatrix},$$

$$H = \begin{pmatrix} \dfrac{1+2i}{3} & \dfrac{\sqrt{2}}{3}(1+i) \\ \dfrac{\sqrt{2}}{3}(-1+i) & \dfrac{1-2i}{3} \end{pmatrix}$$

Example 3. Compute A^{100} for the matrix $A = \begin{pmatrix} \dfrac{3}{2} & \dfrac{1}{2} & \dfrac{1}{2} \\ -\dfrac{1}{4} & \dfrac{3}{4} & \dfrac{1}{4} \\ \dfrac{3}{4} & \dfrac{3}{4} & \dfrac{5}{4} \end{pmatrix}$. The observation to be

made is that the columns of A^{100}, when suitably normalized, tend to the eigenvector of the dominant eigenvalue. The accurate computation of very large or very small numbers is the only significant challenge encountered.

Before computing A^{100}, enter A symbolically, and compute its eigenvalues and eigenvectors in exact arithmentic.

```
> A := matrix(3,3, [3/2,1/2,1/2,-1/4,3/4,1/4,3/4,3/4,5/4]);
```

> q := eigenvects(A);
> A100 := evalm(A^100);

We discover that the dominant eigenvalue is 2 and the associated eigenvector is $\mathbf{v} = \begin{pmatrix} 1 \\ 0 \\ 1 \end{pmatrix}$.

Although the display on the screen is visually unattractive because the rational numbers resulting from this computation contain large integers, Maple has done the computation *exactly* in rational arithmetic. We are looking at the *exact* answer, and there is no round-off error to escape or analyze. The unattractive nature of this exact answer is inherent in the problem and is not an artifact of the computational process.

   We can print the individual columns, which, for this matrix, fit within the available screen width.

> for k from 1 to 3 do col(A100,k); print(); od;

The print statement causes Maple to insert a blank line between columns.

   There is need, however, for caution when converting these exact rational numbers to floating point equivalents. The precision with which this is done determines the acceptability of the result. Converting a matrix to floating point form takes **evalf**, but it must be applied to op(A100) or to evalm(A100). In fact, all functions except **evalf** and **subs** are mapped onto arrays. Both **evalf** and **subs** then require use of **op** (or **evalm**), but when **map** is used, **op** (or **evalm**) is not.

> evalf(op(A100));

The default number of digits used for this calculation is ten. With only ten digits Maple incorrectly displays a matrix M with the structure $M = \begin{pmatrix} \alpha & \alpha & \alpha \\ -\beta & \beta & \beta \\ \alpha & \alpha & \alpha \end{pmatrix}$. The greater precision obtained by

> evalf(op(A100), 31);

shows the middle row to have the pattern $(-\eta, \theta, \eta)$. But we are encouraged to normalize A100 via division by its (1,1)-element.

> q1 := evalm(A100/A100[1,1]);

Rational arithmetic lets us see that the entries in the first and third rows are not exactly the same. Hence,

> q2 := evalf(op(q1));

The matrix q2 strongly suggests that the normalized columns in A100 are nearly the eigenvector **v**. For the very determined, we show how to instruct Maple to round to zero the very small numbers in q2.

> map(fnormal, q2);

Were it not for Maple's ability to implement exact arithmetic and to perform arbitrary precision floating point calculations, we would have to resort to more indirect ways of computing A100. For example, we could diagonalize A and obtain $A = SDS^{-1}$ where S is a matrix whose columns are the eigenvectors of A and D is a diagonal matrix with the eigenvalues of A on the main diagonal. Since $A^2 = (SDS^{-1})(SDS^{-1}) = SD^2S^{-1}$, and since raising the diagonal matrix D to a power results in much less round-off error, we might be able to compute A100 more accurately with lesser tools.

In Lab 9 we saw how tedious the syntax can be for extracting the eigenvalues and eigenvectors from the sequence of lists in q, and for structuring the matrices S and D. We illustrate here a more sophisticated syntax for accomplishing this result. However, we again must avoid the use of D as a name since it is a differentiation operator in Maple.

> f := t -> t[3][1];
> F := t -> t[1];
> S := augment(seq(f(q[k]), k = 1..3));
> d := diag(seq(F(q[k]), k = 1..3));

Clearly, doing the matrix computation $Sd^{100}S^{-1}$ in exact arithmetic is going to produce the exact same result as we obtained for A100. The point of the experiment is to see if this alternative method of computation helps when working in a purely numeric mode. Note the absence of **op** in the computation of Tf since the **inverse** command, applied to S, causes the same evaluation of S^{-1} as **op** normally would.

> Sf := evalf(op(S));
> df := evalf(op(d));
> Tf := evalf(inverse(S));
> A100f := evalm(Sf &* d^100 &* Tf);

If we normalize matrix A100f by dividing all entries by the (1,1)-element, and then using fnormal to round small numbers to zero, we will see that the columns of A100f are nearly the eigenvector **v**.

> q3 := evalm(A100f/A100f[1,1]);
> map(fnormal, q3);

At the default ten digits, we get the same matrix M that we computed by floating A100. The simplest way to increase the number of digits of precision for multiple computations is by setting the default value of the internal variable **Digits** higher. For example, to have

Maple carry out all its numeric computations with 30 digits, execute the following instruction.

> Digits := 30;

To return **Digits** to its default setting of ten, it must be assigned that value.

> Digits := 10;

Even at ten digits of precision we see that the columns of A100f are tending toward a multiple of the eigenvector **v**. Clearly, we can normalize and round down as before. This is left as an exercise in Question 5.

Question 5. Show that at a precision of even ten digits, dividing A^{100} by its (1,1)-element yields a matrix whose columns are approximately the eigenvector corresponding to $\lambda = 2$, the largest eigenvalue of A.

Example 4. Diagonalize $B = \begin{pmatrix} .75 & .25 & .25 \\ -.125 & .375 & .125 \\ .375 & .375 & .625 \end{pmatrix}$ in both floating point and exact

arithmetic, eventually computing B^{100} to observe the effect of B's having $\lambda = 1$ as its largest eigenvalue.

```
> B := matrix(3,3, [.75, .25, .25, -.125, .375, .125, .375, .375, .625]);
> q := eigenvects(B);
> B100 := evalm(B^100);
> f := t -> t[3][1];
> F := t -> t[1];
> S := augment(seq(f(q[k]), k = 1..3));
> d := diag(seq(F(q[k]), k = 1..3));
```

Maple, even in its default mode of ten digits of precision, computes B^{100} in floating point arithmetic and shows that each column tends to the eigenvector of the eigenvalue 1, an eigenvector that again appears to be a multiple of $\mathbf{v} = \begin{pmatrix} 1 \\ 0 \\ 1 \end{pmatrix}$. Moreover, expressing B in terms of the diagonal matrix d does not alter the outcome for B^{100}.

```
> evalm(S &* d^100 &* inverse(S));
```

We also can convert B to a matrix of rational numbers and compute the eigenvectors of B and B^{100} exactly.

```
> Br := map(convert, B, rational);
> eigenvects(Br);
> Br100 := evalm(Br^100);
> evalf(op(Br100));
```

There is no substantial difference in the computation.

Example 5. For $C = \begin{pmatrix} .375 & .125 & .125 \\ -.0625 & .1875 & .0625 \\ .1875 & .1875 & .3125 \end{pmatrix}$, obtain C^{100} and show that its columns

tend toward a multiple of the eigenvector of the dominant eigenvalue. Observe that all eigenvalues of C are less than one in magnitude, so a numeric computation of C^{100}, close to the zero matrix, could be suspect.

```
> C := matrix(3,3, [.375, .125, .125, -.0625, .1875, .0625, .1875, .1875, .3125]);
> eigenve    C);
```

It would appear that the dominant eigenvalue is .5 and that its eigenvector is again the vector **v**. A numeric computation of C^{100}, with normalization, leads to the desired conclusion about its columns.

```
> C100f := evalf(evalm(C^100));
> C100fn := evalm(C100f/C100f[1,1]);
> map(fnormal, C100fn);
```

   For the skeptic worried about the effects of round-off error, we convert C to a rational matrix and repeat the calculation of C^{100}.

```
> Cr := map(convert, C, rational);
> Cr100 := evalm(Cr^100);
> Cr100n := evalm(Cr100/Cr100[1,1]);
> Cr100nf := evalf(op(Cr100n));
> map(fnormal, Cr100nf);
```

The matrix C^{100} is close to the zero matrix, but the results of exact arithmetic and floating point arithmetic are the same.

   Finally, we diagonalize Cr, the rational version of C and again compute C^{100}.

```
> q := eigenvects(Cr);
> f := t -> t[3][1];
> F := t -> t[1];
> S := augment(seq(f(q[k]), k = 1..3));
> d := diag(seq(F(q[k]), k = 1..3));
```

> G := evalm(S &* d^100 &* inverse(S));
> evalm(G - Cr100);

Since the eigenvalues of C are all less than 1 in magnitude, and these are the entries of the diagonal matrix d, raising C to the power 100 raises d to the same power, Hence, d^{100} tends to the zero matrix. Comparing the matrices G and Cr100, we find that in exact arithmetic diagonalizing Cr does not matter.

Question 6. Consider the matrix $T = \begin{pmatrix} \dfrac{1}{2} & \dfrac{1}{4} & \dfrac{1}{2} \\[6pt] \dfrac{1}{3} & \dfrac{1}{2} & \dfrac{1}{2} \\[6pt] \dfrac{1}{6} & \dfrac{1}{4} & 0 \end{pmatrix}$.

(a) Determine the eigenvalues and eigenvectors of T, and then express T as SdS^{-1}.

(b) Compute d^{10}, d^{100}, d^{1000}.

(c) Compute Sd^{10}, Sd^{100}, Sd^{1000}.

(d) Compute S^{-1}.

(e) From the answers to (b) and (c), predict $\lim_{k \to \infty} T^k$.

(f) Check your prediction against a computed value for T^{1000}.

LAB 11

Spectral Theorem

Contents:

Maple Commands:

abs, augment, charpoly, convert/float, diag, dotprod, eigenvects, equal, evalm, fnormal, for/do/od, GramSchmidt, htranspose, inverse, map, matrix, normalize, nullspace, op, seq, solve, sum, transpose, with(linalg), ->

Topics:

Normal matrices, orthogonal and unitary matrices, projection matrices, skew-symmetric and skew-Hermitian matrices, Spectral Theorem, symmetric and Hermitian matrices

Let A be a real normal matrix so that A commutes with its transpose. The Spectral Theorem for A then says that if A is symmetric, then it is orthogonally diagonalizable (A = $S^{-1}DS$, with S an orthogonal matrix and D, diagonal).

If A is a complex normal matrix, it commutes with its hermitian transpose. The Spectral Theorem then says that A is unitarily diagonalizable (A = $U^{-1}DU$, with U a unitary matrix and D, diagonal). Since complex Hermitian, skew-Hermitian, or unitary matrices are all (complex) normal matrices, all these matrices are unitarily diagonalizable also. The important distinction to make is that *all* complex normal matrices are unitarily diagonalizable, whereas only the *symmetric* real normal matrices are orthogonally diagonalizable.

Since Maple's **eigenvects** command does not normalize eigenvectors computed using exact arithmetic, forming unitary or orthogonal transition matrices in these cases will require that we orthonormalize where appropriate.

Example 1. Show that the real symmetric matrix A = $\begin{pmatrix} 1 & 2 & 3 \\ 2 & 2 & 5 \\ 3 & 5 & 1 \end{pmatrix}$ can be orthogonally

diagonalized.

> with(linalg):

```
> A := matrix(3,3, [1,2,3,2,2,5,3,5,1]);
> eigenvects(A);
> eigenvects(A, radical);
```

Both calls to **eigenvects** result in exact, but exceedingly complicated, expressions. The characteristic polynomial for A has roots that really are this complex. It will prove wiser, in this case, to adopt a numeric mode in Maple. We can do this for the eigenvalue-eigenvector calculation by giving A at least one floating point number, or converting A entirely to floats.

```
> Af := map(convert, A, float);
> qf := eigenvects(A);
> f := t -> t[3][1];
> F := t -> t[1];
> S := augment(seq(f(qf[k]), k = 1..3));
> d := diag(seq(F(qf[k]), k = 1..3));
> id := evalm(S &* transpose(S));
> map(fnormal, id, 8);
```

Eigenvectors computed numerically are automatically normalized. Hence, S is an orthogonal matrix, as demonstrated by the computation of S S^t. Round-off error can be accommodated by the **fnormal** command that is mapped onto the matrix S S^t. A display of eight digits shows that id is indeed the identity matrix.

Example 2. The matrix $A = \begin{pmatrix} 0 & 2 & 3 \\ -2 & 0 & 5 \\ -3 & -5 & 0 \end{pmatrix}$ is a real skew-symmetric matrix in which the diagonal entries are perforce zero. The eigenvalues will be pure imaginary, but the characteristic polynomial is a cubic and must have at least one real root. Hence, this real root must be zero.

```
> A := matrix(3,3, [0,2,3,-2,0,5,-3,-5,0]);
> equal(-A, transpose(A));
> q := eigenvects(A, radical);
> S := augment(op(map(normalize,[seq(f(q[k]), k = 1..3)])));
> d := diag(seq(F(q[k]), k = 1..3));
> evalm(S &* htranspose(S));
```

Since the exact representation of the eigenvalues and eigenvectors is not excessively complicated, we have completed the calculation in exact arithmetic. However, that means we have to normalize the eigenvectors before we form the matrix S. The **normalize** command is mapped onto a list of vectors, and the **augment** command acts on a sequence. Hence, the additional **op** command is needed to render the list of normalized vectors as a sequence.

Example 3. The matrix $A = \begin{pmatrix} \frac{1}{\sqrt{2}} & \frac{1}{\sqrt{2}} & 0 \\ \frac{1}{\sqrt{3}} & -\frac{1}{\sqrt{3}} & \frac{1}{\sqrt{3}} \\ \frac{1}{\sqrt{6}} & -\frac{1}{\sqrt{6}} & -\frac{2}{\sqrt{6}} \end{pmatrix}$ is an orthogonal matrix with

eigenvalues that are unit complex numbers. Since one has to be real, it has to be ±1.

```
> s2 := 1/sqrt(2);
> s3 := 1/sqrt(3);
> s6 := 1/sqrt(6);
> A := matrix(3,3 [s2, s2, 0, s3, -s3, s3, s6, -s6, -2*s6]);
> evalm(A &* transpose(A));
> Af := map(convert, A, float);
> q := eigenvects(Af);
> S := augment(seq(f(qf[k]), k = 1..3));
> d := diag(seq(F(qf[k]), k = 1..3));
> evalm(S &* d &* inverse(S));
> evalm(S &* transpose(S));
> map(abs, d);
```

The matrix A is real but not symmetric; it is not orthogonally diagonalizable. It is indeed orthogonal and diagonalizable, but the transition matrix S is not itself orthogonal. Finally, the magnitudes of the eigenvalues are all indeed 1.

Example 4. The matrix $A = \begin{pmatrix} 3 & 4+2i & 2-3i \\ 4-2i & 2 & 1+i \\ 2+3i & 1-i & 5 \end{pmatrix}$ is Hermitian. Its eigenvalues are real,

as are the diagonal entries.

```
> A := matrix(3,3, [3,4+2*I,2-3*I,4-2*I,2,1+I,2+3*I,1-I,5]);
> equal(A, htranspose(A));
> equal(A &* htranspose(A), htranspose(A) &* A);
> Af := map(convert, A, float);
> q := eigenvects(Af);
> S := augment(seq(f(qf[k]), k = 1..3));
> d := diag(seq(F(qf[k]), k = 1..3));
> y := evalm(S &* d &* htranspose(S));
> map(fnormal, y, 2);
> y1 := evalm(S &* htranspose(S));
> map(fnormal, y1, 8);
```

We have tested that A is indeed Hermitian and that it is also normal. The matrix y is

SdS^H, and the **fnormal** command rounds floating point numbers to the desired accuracy, showing that y is actually the matrix Af, the floating point version of A. The matrix y1 is SS^H, and since S is unitary, y1 is the identity. The **fnormal** operator makes it easier to recognize that y1 is indeed the identity.

Example 5. The matrix $A = \begin{pmatrix} 3i & 4+2i & 2-3i \\ -4+2i & 2i & 1+i \\ -2-3i & -1+i & 5i \end{pmatrix}$ is skew-Hermitian, with pure imaginary diagonal elements and eigenvalues.

```
> A := matrix(3,3, [3*I,4+2*I,2-3*I,-4+2*I,2*I,1+I,-2-3*I,-1+I,5*I]);
> equal(-A, htranspose(A));
> equal(A &* htranspose(A), htranspose(A) &* A);
> Af := map(convert, A, float);
> q := eigenvects(Af);
> S := augment(seq(f(qf[k]), k = 1..3));
> d := diag(seq(F(qf[k]), k = 1..3));
> map(fnormal, S);
> map(fnormal, d);
> y := evalm(S &* d &* htranspose(S));
> map(fnormal, y, 2);
> y1 := evalm(S &* htranspose(S));
> map(fnormal, y1, 8);
```

First, we have verified that A is skew-Hermitian, and that it is normal. Then, after computing the eigenvalues and eigenvectors in floating point arithmetic, we have verified that A was unitarily diagonalizable: SdS^H is A, and SS^H is the identity.

Example 6. The matrix $A = \begin{pmatrix} \dfrac{1-i}{2} & \dfrac{1+i}{3} & -\dfrac{1+3i}{6} \\ \dfrac{1-i}{2} & -\dfrac{1+i}{3} & \dfrac{1+3i}{6} \\ 0 & \dfrac{2-i}{3} & \dfrac{2}{3} \end{pmatrix}$ is unitary and therefore has complex numbers of magnitude 1 for eigenvalues.

```
> A := matrix(3,3, [(1-I)/2,(1+I)/3,-(1+3*I)/6,(1-I)/2,-(1+I)/3,(1+3*I)/6,0,(2-I)/3,2/3]);
> evalm(A &* htranspose(A));
> equal(A &* htranspose(A), htranspose(A) &* A);
> Af := map(convert, A, float);
> q := eigenvects(Af);
> S := augment(seq(f(qf[k]), k = 1..3));
> d := diag(seq(F(qf[k]), k = 1..3));
```

```
> y := evalm(S &* d &* htranspose(S));
> map(fnormal, y, 8);
> y1 := evalm(S &* htranspose(S));
> map(fnormal, y1, 8);
> map(abs, d);
```

First, we check that A is unitary by showing that AA^H is the identity; then we check that A is normal. We compute the eigenvalues and eigenvectors in floating point arithmetic and check that A was unitarily diagonalizable, recalling that all complex normal matrices are unitarily diagonalizable. We further verify that SS^H is the identity. Finally, we see that the eigenvalues all have magnitude 1.

Example 7. The matrix $B = AdA^H$, where A is the matrix of Example 6 and d is the diagonal matrix whose diagonal entries are (2, 3+i, 3-i). From Example 6 we know that A is unitary, so B is unitarily similar to a diagonal matrix. B is normal, but not Hermitian, skew-Hermitian, or unitary. Its eigenvalues are neither all real, nor pure imaginary unit complex numbers.

```
> d := diag(2, 3+I, 3-I);
> B := evalm(A &* d &* htranspose(A));
> equal(B &* htranspose(B), htranspose(B) &* B);
> equal(B, htranspose(B));
> equal(-B, htranspose(B));
> evalm(B &* htranspose(B));
```

The matrix B is normal since $BB^H = B^HB$, but B is not Hermitian since $B \neq B^H$. Neither is B skew-Hermitian since $B^H \neq -B$. And finally, B is not unitary since BB^H is not the identity.

Question 1. Construct an example of a real matrix A which is normal ($AA^t = A^tA$) but is neither symmetric, skew-symmetric, nor orthogonal.

For modest-sized matrices where the arithmetic is minimal, we can explore the steps of the algorithm that orthogonally or unitarily diagonalizes a matrix. For symmetric real matrices and Hermitian complex ones, the eigenvectors corresponding to distinct eigenvalues are necessarily orthogonal. When such matrices have eigenvalues with algebraic multiplicity greater than one, the basis for the eigenspace of that repeated eigenvalue may have to be orthogonalized by the Gram-Schmidt process. It is always instructive to work through several examples when contrasting the behavior of definitive cases.

Example 8. The matrix $A = \begin{pmatrix} 13 & -2 & -5 \\ -2 & 10 & -2 \\ -5 & -2 & 13 \end{pmatrix}$ is real and symmetric; hence, it is orthogonally diagonalizable. Find S, the orthogonal transition matrix defining the similarity transformation which diagonalizes A.

```
> A := matrix(3,3, [13,-2,-5,-2,10,-2,-5,-2,13]);
> q := eigenvects(A);
> for k from 1 to 3 do v.k := f(q[k]); od;
> dotprod(v1, v2); dotprod(v1,v3); dotprod(v2,v3);
> S := augment(op(map(normalize,[v.(1..3)]))));
> d := diag(seq(F(q[k]),k=1..3));
> evalm(S &* transpose(S));
> evalm(S &* d &* transpose(S));
```

The eigenvalues of A are distinct, and the eigenvectors are given simply with integer components. Hence we work symbolically and isolate these eigenvectors, showing that **v1**, **v2**, and **v3** are indeed mutually orthogonal. Then, normalizing the eigenvectors we form S with the normalized eigenvectors as its columns, and form the diagonal matrix d with the eigenvalues on the main diagonal. Finally, we demonstrate that S so formed is an orthogonal matrix and that SdS^t is A.

Example 9. The matrix $A = \begin{pmatrix} 4 & -2 & -2 \\ -2 & 4 & -2 \\ -2 & -2 & 4 \end{pmatrix}$ is real and symmetric, and has repeated eigenvalues. We again determine the similarity transformation that orthogonally diagonalizes A.

```
> A := matrix(3,3, [4,-2,-2,-2,4,-2,-2,-2,4]);
> q := eigenvects(A);
> v1 := q[1][3][1];
> v2 := q[1][3][2];
> v3 := q[2][3][1];
> dotprod(v1,v2); dotprod(v1, v3); dotprod(v2, v3);
> qq := GramSchmidt([v1,v2]);
> qqq := [op(qq), op(v3)];
> S := augment(op(map(normalize,qqq)));
> d := diag(q[1][1], q[1][1], q[2][1]);
> evalm(S &* transpose(S));
> evalm(S &* d &* transpose(S));
```

Indeed, A has repeated eigenvalues, but the eigenvectors admit integer arithmetic. We therefore proceed symbolically, extracting the eigenvectors from the sequence of lists in q.

The function f used earlier is not sophisticated enough to detect the repeated eigenvalue or the two members of the basis of the eigenspace associated with the eigenvalue 6. Our intervention requires that we observe which of the two lists in q contain the repeated eigenvalue. There is no canonical ordering for such returns, so reexecuting the **eigenvects** command could cause the two lists in the sequence q to be interchanged.

Once the eigenvectors **v1**, **v2**, and **v3** are available, we demonstrate that indeed, **v1** and **v2**, the eigenvectors of the repeated eigenvalue 6, are not orthogonal, but they are both orthogonal to **v3**, the eigenvector from the eigenvalue 0. These two eigenvectors are orthogonalized by the Gram-Schmidt process, and then all three candidates for columns of S are normalized. Finally, S is formed and d constructed. Again, the function F is not sophisticated enough to detect the repeated eigenvalues, so obtaining d is a bit more tedious. Finally we show that S is orthogonal and that SdS^t is really A.

Question 2. As in Example 8, show that $A = \begin{pmatrix} 0 & \frac{1}{\sqrt{3}} & \frac{2}{\sqrt{6}} \\ \frac{1}{\sqrt{3}} & 0 & 0 \\ \frac{2}{\sqrt{6}} & 0 & 0 \end{pmatrix}$ is orthogonally

diagonalizable by constructing the transition matrix S and the diagonal matrix d. Verify that S is indeed orthogonal and that $SdS^t = A$.

The Spectral Theorem states that a matrix A can be written as a linear combination of projection matrices; hence, $A = \sum_k \lambda_k P_k$, where the λ_k are the eigenvalues of A, and P_k projects onto the eigenspace of λ_k. If the columns vectors in the matrix M are independent, then the matrix P that projects onto the column space of M is given by $P = M(M^tM)^{-1}M^t$. If the columns of M are orthonormal (and therefore necessarily independent), then P is obtained as MM^t.

Example 10. Write the matrix in Example 8 as a linear combination of projection matrices.

```
> A := matrix(3,3, [13,-2,-5,-2,10,-2,-5,-2,13]);
> q := eigenvects(A);
> for k from 1 to 3 do v.k := normalize(f(q[k])); od;
> for k from 1 to 3 do P.k := evalm(v.k &* transpose(v.k)); od;
> for k from 1 to 3 do equal(P.k, P.k^2); od;
> A1 := evalm(sum('F(q[k])*P.k', 'k' = 1..3));
```

Since each eigenspace for A is one-dimensional, a matrix M from which a projection matrix P will be constructed contains only a single vector. Moreover, the characteristic of a projection matrix P being $P^2 = P$, we easily demonstrate that we have found projection matrices. And finally, we observe that the matrix A1 matches the matrix A.

The alternative computation in which the eigenvectors are not normalized can also be implemented in Maple. However, for this case of projecting onto the space spanned by a single vector **u**, the product **uu**^{t} is a scalar which Maple's **inverse** command rejects, since it is looking for a matrix and not a scalar. Hence, the vector **u** would need to be converted to a matrix, since Maple distinguishes between a 1x1 matrix and a scalar. The following syntax would therefore be needed in Maple, and we remind the reader that entering a multi-line input is platform dependent. On a Macintosh, use the **Return** key instead of the **Enter** key. On other Windowed platforms (PC or UNIX), use the **Shift-Enter** keys simultaneously to obtain a new line without sending the input off to Maple for computation.

```
> for k from 1 to 3 do m.k := convert(f(q[k]), matrix); od;
> for k from 1 to 3 do
     P.k := evalm(m.k &* inverse(transpose(m.k) &* m.k) &* transpose(m.k));
     od;
```

Question 3. Write the matrix from Question 2 as a linear combination of projection matrices.

Example 11. Write the matrix $A = \begin{pmatrix} 4 & 1 & 1 \\ 1 & 4 & 1 \\ 1 & 1 & 4 \end{pmatrix}$, which has repeated eigenvalues, as a linear combination of projection matrices.

```
> A := matrix(3,3, [4,1,1,1,4,1,1,1,4]);
> cp := charpoly(A, x);
> q := solve(cp = 0, x);
> q1 := nullspace(A - 3);
> q2 := nullspace(A - 6);
> q3 := map(normalize, GramSchmidt(q1));
> q4 := augment(op(q3));
> P3 := evalm(q4 &* transpose(q4));
> q5 := normalize(q2[1]);
> P6 := evalm(q5 &* transpose(q5));
> A1 := evalm(3*P3 + 6*P6);
```

There are only two projection matrices, one projecting onto the two-dimensional eigenspace of $\lambda = 3$ and one projecting onto the one-dimensional eigenspace of $\lambda = 6$. We have obtained the eigenvalues and eigenvectors by implementing computations that would be the norm for working with a pencil and paper. The results are the same as produced by a call to **eigenvects**. However, with the latter approach, one needs to be careful extracting the eigenvalues and eigenvectors from the sequence of lists obtained, since there is now an eigenvalue with algebraic and geometric multiplicity two.

Question 4. Write the matrix $A = \begin{pmatrix} 2 & 2 & -1 \\ 2 & -1 & 2 \\ -1 & 2 & 2 \end{pmatrix}$ as a linear combination of projection matrices. As in the last example, implement steps conformable to a pencil-and-paper mode of computation. Then check your results with a call to **eigenvects**.

LAB 12

Applications of the Spectral Theorem

Contents:
Maple Commands:

angle, augment, collect, convert/list, cos, diag, display, eigenvects, equal, evalf, evalm, expand, extend, fnormal, for/do/od, Gramschmidt, hessian, if/then/else/fi, Im, implicitplot, inverse, lhs, map, matrix, normalize, nullspace, op, plot, print, radnormal, rank, rationalize, Re, readlib, seq, simplify, sin, singularvals, sqrt, subs, Svd, transpose, vector, with(linalg), with(plots), ->

Topics:

Quadrics and their graphs, rotations and reflections, singular value decomposition and pseudoinverse, Spectral Theorem for orthogonal matrices

Maple has the function **Svd** for computing a singular value decomposition of a matrix A. However, Maple does not have a command for computing the pseudoinverse of A, so we will show how to obtain the pseudoinverse from the output of **Svd**. This command is in the resident Maple library, not in the *linalg* package, but we will need some matrix manipulations almost immediately, so we begin by loading the *linalg* package.

Example 1. Obtain the singular value decomposition of

$$A = \begin{pmatrix} 2 & 3 & 4 \\ 1 & -1 & 3 \\ 3 & 2 & 7 \\ 1 & 4 & 1 \end{pmatrix}$$

and use it to find A^+, the pseudoinverse of A. Form the products $C = A^+A$ and $G = AA^+$.

```
> with(linalg):
> A := matrix(4,3, [2,3,4,1,-1,3,3,2,7,1,4,1]);
> svf := evalf(Svd(A, U, V));
> SIG := map(fnormal,extend(diag(op(convert(svf,list))),1,0,0));
> AA := evalm(U &* SIG &* transpose(V));
```

```
> SIGPLUS := map(t -> if t = 0 then t else 1/t fi, transpose(SIG));
> APLUS := evalm(V &* SIGPLUS &* transpose(U));
> C := evalm(APLUS &* A);
> G := evalm(A &* APLUS);
```

The **Svf** command does no computation without the accompanying **evalf**. The output is then a vector of singular values. These singular values must be placed in the "diagonal" matrix Σ. We do this in stages. First, we convert svf, the vector of singular values, to a list, and then we extract these values from the list to present a sequence to the **diag** command. This forms a true diagonal matrix which must be extended with one more row, but no more columns, of zeros. This matrix is Σ, and we take one more liberty of rounding exceedingly small singular values to zero by applying the **fnormal** command. This, then, is the matrix Σ, called SIG in the example.

Maple's **Svd** command also computes and assigns to the names U and V the matrices Q_1 and Q_2, respectively. Hence, we can immediately check that $Q_1 \Sigma Q_2^t$, the singular value decomposition, is equivalent to A.

To form Σ^+ from Σ, we replace the nonzero elements in Σ with their reciprocals and then transpose the resulting matrix. The replacement by reciprocals is accomplished in Maple by mapping an appropriate function onto the elements of SIG. The function simply tests each element to see which ones are nonzero and then reciprocates each such element. The transpose is actually taken first.

The pseudoinverse is obtainable as $A^+ = Q_2 \Sigma^+ Q_1^t$ and this is computed in Maple as APLUS.

Question 1. Explain what multiplication by the matrix C in Example 1 will do to vectors in the row space of A. Find a basis for the row space and check your contention. Explain what multiplication by the matrix C will do to vectors in the null space of A. Find a basis for the null space and test your assertion.

Question 2. Explain what multiplication by the matrix G of Example 1 will do to vectors in the column space of A. Find a basis for the column space and test your claim. Explain what multiplication by G does to vectors in the null space of A^t. Using a basis for the null space of A^t, verify your claim.

It is instructive to obtain the singular value decomposition and pseudoinverse of Example 1 by manipulating the eigenvalues and eigenvectors of the matrix A^tA. This computation can actually be done in exact arithmetic in Maple. A list of exact singular values are obtained in Maple via the **singularvals** command in the *linalg* package. However, no advantage accrues, since these same values will arise from the **eigenvects** command mandated by the constraints of this next example.

Example 2. Using the eigenvalues and eigenvectors of A^tA, obtain the singular value decomposition and the pseudoinverse of the matrix A in Example 1.

```
> rank(A);
> singularvals(A);
> AtA := evalm(transpose(A) &* A);
> rank(AtA);
> q := eigenvects(AtA, radical);
> f := t -> t[3][1];
> F := t -> t[1];
```

We need to extract the eigenvalues and eigenvectors, but in a certain order. By inspection, we determine that order to be 2,3 and place this information in a list. However, there is no canonical ordering for this information, and upon reexecution of the command this ordering can be different. The user must intervene and observe the desired ordering. Maple's **seq** command can iterate over the list produced.

```
> ordering_list := [2,3];
> ordered_singularvals := seq(sqrt(F(q[k])), k = ordering_list);
> SIGr := diag(ordered_singularvals);
> basis_RowSp_A := [seq(f(q[k]), k = ordering_list)];
```

We now have a basis for the row space of A, a basis consisting of eigenvectors of A^tA. However, this basis is yet to be orthonormalized.

```
> Q2r := augment(op(map(normalize, GramSchmidt(basis_RowSp_A))));
> evalf(op(Q2r));
```

The appropriate orthonormal basis for the column space of A, a basis consisting of eigenvectors of A^tA, and satisfying $\mathbf{u}_k = \dfrac{1}{\sigma_k} A\mathbf{v}_k$, where σ_k is the kth singular value and $\mathbf{v}_k$ is the kth column of Q_{2r}, is obtained by the following matrix product that puts the basis vectors $\mathbf{u}_k$ into the matrix Q_{1r} as columns.

```
> Q1r := evalm(A &* Q2r &* inverse(SIGr));
> evalf(op(Q1r));
```

We test that $Q_{1r}\,\Sigma_r\,Q_{2r}^t$ suffices as a decomposition of A.

```
> is_a := evalm(Q1r &* SIGr &* transpose(Q2r));
> evalf(op(is_a));
```

We can obtain A exactly by applying the undocumented **radnormal** command to each element of the matrix is_a. Although this command is in Maple's miscellaneous library and is available through a **readlib**, it fails to respond to any call to the help system.

```
> readlib(radnormal);
```

```
> map(radnormal, is_a);
```

To obtain the complete singular value decomposition in terms of $Q_1 \Sigma Q_2^t$, we continue, first finding an orthonormal basis for the null space of A, a basis chosen from the eigenvectors of A^tA. This will yield Q_{2n} and hence, Q_2. We again caution that the ordering of the lists in the sequence q can be different for different executions of the command. An inspection should be made of the particular ordering your session has generated.

```
> v3 := q[1][3][1];
> Q2n := normalize(v3);
> Q2 := augment(Q2r, Q2n);
> Q2f := evalf(op(Q2));
```

An orthonormal basis for the null space of A^t gives the columns of Q_{1n} and therefore we will have Q_1.

```
> basis_N_At := nullspace(transpose(A));
> Q1n := augment(op(map(normalize, GramSchmidt(basis_N_At))));
> Q1 := augment(Q1r, Q1n);
> Q1f := evalf(op(Q1));
```

The full matrix of singular values, Σ, requires that we fill out Σ_r with an appropriate number of zeros.

```
> SIG := extend(SIGr, 2, 1, 0);
```

We recover A from the full singular value decomposition. The matrix so formed is the same as is_a, and it yields to the same simplification.

```
> is_A := evalm(Q1 &* SIG &* transpose(Q2));
> map(radnormal, is_A);
```

The pseudoinverse A^+ can be calculated from Q_{1r}, Σ_r, and Q_{2r}. Working in exact arithmetic, we will need the same simplification that was used on is_a and is_A.

```
> aplus := evalm(Q2r &* transpose(1/SIGr) &* transpose(Q1r));
> map(radnormal, aplus);
> evalf(op(aplus));
```

Having obtained A^+ by the curtailed singular value decomposition, we show that we get the same pseudoinverse from the full decomposition.

```
> SIGplus := map(t -> if t = 0 then t else 1/t fi, transpose(SIG));
> Aplus := evalm(Q2 &* SIGplus &* transpose(Q1));
> map(radnormal, Aplus);
> evalf(op(Aplus));
```

Thus, we have shown that we can compute A^+ in exact arithmetic, and have therefore made it simpler to discuss the projection properties of the matrices A^+A and AA^+. Finally, we compare the floating point equivalents of our exact decomposition with the floating point decomposition computed numerically by Maple.

```
> print(U, Q1f);
> print(V, Q2f);
```

Question 3. Using the method of Example 2, find the singular value decomposition and pseudoinverse for $C = \begin{pmatrix} 1 & 3 & -1 & 2 \\ 1 & 2 & 1 & 0 \\ 1 & 4 & -3 & 4 \\ 3 & 7 & 1 & 2 \end{pmatrix}$. Then check your results by the numeric method in Example 1.

Question 4. Using the numeric method of Example 1, find the singular value decomposition and the pseudoinverse for the matrix $M = C^t$, where C is the matrix in Question 3. How are the answers for the singular value decompositions related? By checking other examples, formulate a conjecture. Prove your conjecture. How are the answers for the pseudoinverses related? By checking other examples, formulate a conjecture. Prove your conjecture.

Section V.4 of Terry Lawson's <u>Linear Algebra</u> text gives a normal form R for an orthogonal matrix A. In this normal form $A = ORO^t$, where O is a special orthogonal matrix and R has 2x2 blocks corresponding to rotations. The 2x2 blocks correspond to two-dimensional subspaces, so that multiplication by A gives a rotation of this subspace through an angle θ. The angle is related to a complex eigenvalue $\cos(\theta) + i\sin(\theta)$. The basis $\mathbf{w}$, $\mathbf{v}$ for the subspace comes from the complex eigenvector $\mathbf{v} + i\mathbf{w}$ for the eigenvalue $\cos(\theta) + i\sin(\theta)$. In the 3x3 case with a complex eigenvalue, and determinant 1, the eigenvector for the eigenvalue 1 gives the axis of rotation.

Example 3. Factor the orthogonal matrix $A = \begin{pmatrix} \dfrac{1+\sqrt{3}}{3} & \dfrac{1}{3} & \dfrac{1-\sqrt{3}}{3} \\[2mm] \dfrac{1-\sqrt{3}}{3} & \dfrac{1+\sqrt{3}}{3} & \dfrac{1}{3} \\[2mm] \dfrac{1}{3} & \dfrac{1-\sqrt{3}}{3} & \dfrac{1+\sqrt{3}}{3} \end{pmatrix}$ into SRS^t,

where S is itself an orthogonal matrix and R is a rotation in a plane in R^3. We begin by finding the eigenvalues and eigenvectors of A.

```
> s1 := (1 + sqrt(3))/3;
> s2 := (1 - sqrt(3))/3;
> A := matrix(3,3, [s1, 1/3, s2, s2, s1, 1/3, 1/3, s2, s1]);
> q := eigenvects(A);
> for k from 1 to 3 do v.k := f(q[k]); od;
> for k from 1 to 3 do m.k := F(q[k]); od;
> u := map(Re, v1);
> v := map(Im, v1);
> t := argument(m1);
> S := augment(op(map(normalize, [v3, v, u])));
> equal(inverse(S), transpose(S));
> R := matrix(3,3, [1,0,0,0,cos(t),-sin(t),0,sin(t),cos(t)]);
> A1 := evalm(S &* R &* transpose(S));
```

The vectors **u** and **v** are the real and imaginary parts of the first complex eigenvector. The argument of the complex eigenvalue m1 is represented by t. The orthogonal matrix S consists of columns that are the normalized versions of **u**, **v**, and **v3**, the eigenvector corresponding to the real eigenvalue $\lambda = 1$. A simple test indicates that S is indeed an orthogonal matrix. The matrix R is a rotation in the plane spanned by the vectors **u** and **v**. The matrix $A1 = SRS^t$ is indeed the original matrix A.

Question 5. As in Example 3, factor each of the following matrices into the form SRS^t, where S is an orthogonal matrix and R is a rotation in a plane spanned by two of the eigenvectors.

(a) $\begin{pmatrix} \dfrac{1}{\sqrt{3}} & \dfrac{1}{\sqrt{2}} & \dfrac{1}{\sqrt{6}} \\[2mm] \dfrac{1}{\sqrt{3}} & 0 & -\dfrac{2}{\sqrt{6}} \\[2mm] \dfrac{1}{\sqrt{3}} & -\dfrac{1}{\sqrt{2}} & \dfrac{1}{\sqrt{6}} \end{pmatrix}$; (b) $\begin{pmatrix} -\dfrac{1}{\sqrt{3}} & \dfrac{1}{\sqrt{2}} & \dfrac{1}{\sqrt{6}} \\[2mm] -\dfrac{1}{\sqrt{3}} & 0 & -\dfrac{2}{\sqrt{6}} \\[2mm] -\dfrac{1}{\sqrt{3}} & -\dfrac{1}{\sqrt{2}} & \dfrac{1}{\sqrt{6}} \end{pmatrix}$.

We next consider the problem of analyzing the equation of a conic section. Plotting the conic is easily handled by Maple's **implicitplot** command in the *plots* package. A more challenging task is finding a coordinate system in which the conic is in standard form. This typically involves a rotation of coordinates and a translation to a new origin.

Example 4. Plot the rotated conic represented by the equation

$$x^2 + 3xy + y^2 = 1.$$

```
> with(plots);
> q := x^2 + 3*x*y + y^2 = 1;
> implicitplot(q, x = -3..3, y = -3..3);
```

The graph of a rotated hyperbola results. The typical interface to Maple V Release 3 has interactive controls on the plot window for setting the aspect ratio to 1-to-1, thereby making circles look like circles (instead of ellipses). This functionality is also available from the keyboard.

```
> implicitplot(q, x = -3..3, y = -3..3, scaling = constrained);
```

The lefthand side of the given equation can be represented as the quadratic form

$$(x \quad y) \begin{pmatrix} 1 & \frac{3}{2} \\ \frac{3}{2} & 1 \end{pmatrix} \begin{pmatrix} x \\ y \end{pmatrix}.$$

The coefficient matrix is actually $\frac{1}{2}$ the matrix of second partial derivatives of the lefthand side of the given equation. We can obtain this in Maple by the **hessian** command in the *linalg* package.

```
> A := evalm(hessian(lhs(q), [x,y])/2);
```

The major and minor axes of this rotated conic are related to the eigenvectors of the quadratic form whose matrix is the symmetric matrix A. Caution is again needed when interpreting the ordering of the lists in q1. If Maple reverses the lists in q1, the vectors **v1** and **v2** will be interchanged. We will have no trouble with computing the angle between **v1** and **v2**, but shortly, when we seek the angle between these eigenvectors and the coordinate axes, any such reversal of ordering will matter.

```
> q1 := eigenvects(A);
> v1 := f(q1[1]);
> v2 := f(q1[2]);
> angle(v1, v2);
```

This tells us that the angle between the eigenvectors is $\frac{\pi}{2}$, a fact we already knew by the orthogonality of the eigenvectors of a real symmetric matrix. We really want to know the angle between the eigenvectors and the x-axis.

```
> e1 := vector([1,0]);
> angle(v1, e1);
> angle(v2, e1);
```

We readily discover that **v1** makes an angle of $\frac{3\pi}{4}$, and **v2**, an angle of $\frac{\pi}{4}$, with the x-axis. Hence, a rotation of $\frac{\pi}{4}$ would bring the conic to standard form.

```
> t := Pi/4;
> q2 := {x = u*cos(t) - v*sin(t), y = u*sin(t) + v*cos(t)};
> q3 := expand(subs(q2, q));
```

Equation q3, $\frac{5}{2}u^2 - \frac{1}{2}v^2 = 1$, now gives the rotated hyperbola, but in the uv-coordinate system. But we can plot both images of this hyperbola and even use color to distinguish each.

```
> p1 := implicitplot(q3, u = -3..3, v = -3..3, color = black):
> p2 := implicitplot(q, x = -3..3, y = -3..3, color = red):
> p3 := plot({[x, x, x = -3..3], [x, -x, x = -3..3]}, color = red):
> display([p1, p2, p3], scaling = constrained);
```

The rotated hyperbola appears in red, along with its axes, also in red. The hyperbola in standard form appears in black, along with its (default) axes in black. Assigning a plot data structure to a variable allows the structure to be manipulated, and joined to other plot data structures via the **display** command.

Example 5. Plot the rotated conic represented by the equation

$$x^2 + 4xy + 5y^2 = 1.$$

Written as a quadratic form, we have $(x \quad y)\begin{pmatrix} 1 & 2 \\ 2 & 5 \end{pmatrix}\begin{pmatrix} x \\ y \end{pmatrix}$. We do the computations in Maple, eventually finding the eigenvalues and eigenvectors.

```
> q := x^2 + 4*x*y + 5*y^2 = 1;
> implicitplot(q, x = -3..3, y = -1..1, numpoints = 1000);
> A := evalm(hessian(lhs(q), [x,y])/2);
> q1 := eigenvects(A, radical);
```

```
> v1 := f(q1[1]);
> v2 := f(q1[2]);
> t1 := angle(v1, e1);
> t2 := angle(v2, e1);
```

The conic is a rotated ellipse. The additional argument to **implicitplot** causes Maple to use a minimum of 1000 points in its plot algorithm, making for a smoother graph. The additional argument in the **eigenvects** command causes Maple to give the eigenvalues and eigenvectors in radical notation rather than in its *RootOf* notation. The angle between the eigenvectors **v1** and **v2** is difficult, but not impossible, to obtain in exact form. First, evaluate the angles t1 and t2 in degrees.

```
> evalf(t1*180/Pi);
> evalf(t2*180/Pi);
```

It appears that t1 is 67.5 degrees and t2, 157.5 degrees. Hence, **v2** makes an angle of 90 degrees with **v1**, an obvious conclusion given the symmetry of the matrix A. However, 67.5 degrees is equivalent to $\frac{\pi}{4} + \frac{\pi}{8} = \frac{3\pi}{8}$. Now

```
> simplify(cos(3*Pi/8));
```

yields $\frac{1}{2}\sqrt{2-\sqrt{2}}$ so the argument of the arccosine function in t1 must reduce to this expression. It does, but Maple does not readily provide this transformation.

```
> readlib(rationalize);
> sqrt(expand(rationalize(radnormal(cos(t1)^2))));
```

Consequently, we will rotate the conic through an angle of $\frac{3\pi}{8}$.

```
> t := 3*Pi/8;
> q2 := {x = u*cos(t) - v*sin(t), y = u*sin(t) + v*cos(t)};
> q3 := subs(q2, q);
> q4 := radnormal(q3);
> q5 := collect(expand(q4), [u,v]);
```

Equation q5 contains the original rotated ellipse of equation q, but now in the standard form

$$(3 + 2\sqrt{2})u^2 + (3 - 2\sqrt{2})v^2 = 1$$

in the uv-coordinate system. We create a plot showing both the rotated and the unrotated ellipses.

```
> p1 := implicitplot(q, x = -3..3, y = -1..1, numpoints = 1000, color = red):
> p2 := plot({tan(t)*x, -cot(t)*x}, x = -3..3, color = red):
> p3 := implicitplot(q5, u = -1..1, v = -3..3, numpoints = 1000, color = black);
> display([p1, p2, p3], scaling = constrained, view = [-3..3, -3..3]);
```

Question 6. For each of the following equations, use the eigenvalue-eigenvector structure of the corresponding matrix of the associated quadratic form to determine by what angle one must rotate the axes to put the graph in standard form. Then sketch the graph.

(a) $3x^2 - 2xy + y^2 = 1$

(b) $3x^2 - 4xy + y^2 = 1$.

LAB 13

Normal Forms

Contents:

Maple Commands: *augment, charpoly, convert/float, det, diag, eigenvects, evalm, factor, fnormal, for/do/od, if/then/else/fi, inverse, jordan, map, matrix, minpoly, normalize, nullspace, op, pivot, radsimp, seq, submatrix, subs, subvector, swaprow, transpose, vector, with(linalg), ->, $*

Topics: Characteristic polynomial, Jordan block, Jordan canonical form, Jordan string, minimal polynomial, normal forms for quadratic forms

We have learned how to orthogonally diagonalize a symmetric matrix A by an appropriate similarity transform of the form $A = SDS^t$. We next consider putting such a matrix A into the normal form $TD(p,n,d)T^t$, a process that involves reordering the eigenpairs before normalizing.

Example 1. If $A = \begin{pmatrix} -2 & -7 & 1 \\ -7 & -2 & -1 \\ 1 & -1 & 4 \end{pmatrix}$, find the similarity transformation that orthogonally

diagonalizes A to the normal form $TD(p,n,d)T^t$, where $D(p,n,d)$ is a diagonal matrix whose entries are a sequence of p 1's, n negative 1's and d zeros. We begin by finding the eigenvalues and eigenvectors, ordering the eigenvalues to correspond with the desired structure in $D(p,n,d)$, ordering the eigenvectors accordingly, normalizing the eigenvectors, and placing them in a matrix S_1. The diagonal matrix R whose entries are the ordered eigenvalues is then factored as $R = S_2D(p,n,d)S_2$. The transition matrix for the required similarity transform is then $S = S_1S_2$.

It behooves the reader to be wary of Maple's propensity to give the sequence of lists in q, a result of the **eigenvects** command, in different orders. An inspection of the output needs to be made each time the command is executed, so that the ordering can be specific to the sequence provided by Maple at the time of execution.

```
> with(linalg):
> A := matrix(3,3, [-2,-7,1,-7,-2,-1,1,-1,4]);
> q := eigenvects(A);
> f := t -> t[3][1];
> F := t -> t[1];
> ordering_list := [3,1,2];
> S1 := augment(op(map(normalize,[seq(f(q[k]), k = ordering_list)]))));
> R := diag(seq(F(q[k]), k = ordering_list));
> g := t -> if t > 0 then 1/sqrt(t) elif t < 0 then -1/sqrt(-t) else 0; fi;
> S2 := map(g, R);
> S := map(radsimp, evalm(S1 &* S2));
> evalm(transpose(S) &* A &* S);
```

Not every matrix yields to exact arithmetic. Even 3x3 matrices can easily generate large and complicated expressions for the exact representations of eigenvalues and eigenvectors. Hence, even in Maple we need facility with floating point arithmetic.

Example 2. Work Example 1 in floating point arithmetic. By now, it should go without saying that the ordering of the lists in q1, the output of the **eigenvects** command, can be different each time the command is executed.

```
> A1 := map(convert, A, float);
> q1 := eigenvects(A1);
> ordering_list := [1,2,3];
> S1 := augment(seq(f(q1[k]), k = ordering_list));
> R := diag(seq(F(q1[k]), k = ordering_list));
> S2 := map(g, R);
> S := evalm(S1 &* S2);
> isD := evalm(transpose(S) &* A1 &* S);
> map(fnormal, isD, 8);
```

Question 1. Use the methods of Example 1 or 2 (as appropriate) to put the following symmetric matrices into the normal form $S^t A S = D(p,n,d)$ by finding the transition matrix S. Verify that your matrix S accomplishes its task.

$$(a) \begin{pmatrix} 1 & 1 & 2 \\ 1 & 2 & 1 \\ 2 & 1 & 3 \end{pmatrix}; \quad (b) \begin{pmatrix} 1 & 2 & 3 & 4 \\ 2 & 1 & 4 & 3 \\ 3 & 4 & 1 & 2 \\ 4 & 3 & 2 & 1 \end{pmatrix}.$$

Example 3. For the matrix of Examples 1 and 2, obtain the same normal form by the method of symmetric Gaussian elimination.

```
> id := diag(1$3);
> AI := augment(A, id);
```

```
> AI1 := pivot(AI, 1, 1);
> P1t := submatrix(AI1, 1..3, 4..6);
> AI2 := evalm(P1t &* A &* transpose(P1t));
> AI3 := augment(AI2, id);
> AI4 := pivot(AI3, 2, 2);
> P2t := submatrix(AI4, 1..3, 4..6);
> AI5 := evalm(P2t &* AI2 &* transpose(P2t));
> P3t := swaprow(id, 1, 3);
> AI6 := evalm(P3t &* AI5 &* transpose(P3t));
> Tt := diag(1/sqrt(18/5), 1/sqrt(45/2), 1/sqrt(2));
> S := transpose(Tt &* P3t &* P2t &* P1t);
> d := evalm(transpose(S) &* A &* S);
```

The **pivot** command applies Gaussian elimination to all the elements of a column, creating zeros both above the pivot element and below. The matrix P1t captures these Gaussian elimination steps, and AI2 is the result of applying the same steps both to the rows of A and to the columns. Hence, P2t captures the reductions applied to AI2 and AI5 reflects the performance of these operations on the rows and columns of AI2. The matrix AI5 is diagonal, but the (1,1)-element is negative. P3t is therefore a permutation matrix that will be applied to AI5 to give the correct ordering to the diagonal elements, as shown in AI6. Tt is the diagonal scaling matrix that ensures the nonzero diagonal elements will have magnitude 1, and S captures all the row reduction steps. Hence, d is the desired canonical form for A.

Question 2. Use the method of Example 3 to rework Question 1, finding the matrix S for which $S^t A S = D(p,n,d)$. Verify that your matrix S does what it should.

We turn our attention now to the Jordan canonical form, found, for example, in Section 6.2 of Terry Lawson's <u>Linear Algebra</u>. Recall that the Cayley-Hamilton theorem claims a matrix A satisfies its characteristic equation, the equation formed by setting its characteristic polynomial equal to zero. In the next example we will verify this claim about A and its characteristic polynomial, then find the minimal polynomial for A, and finally, find the Jordan canonical form for A.

Example 4. Let $A = \begin{pmatrix} 1 & 0 & 1 & 0 \\ -\frac{1}{2} & 1 & 0 & \frac{1}{2} \\ 0 & 0 & 1 & 0 \\ 0 & 0 & 1 & 1 \end{pmatrix}$. Obtain the characteristic polynomial and verify that A satisfies its characteristic equation.

```
> A := matrix(4,4, [1,0,1,0,-1/2,1,0,1/2,0,0,1,0,0,0,1,1]);
> cp := charpoly(A, x);
```

```
> det(A - x);
> evalm(subs(x = A, cp));
```

We have given two options for finding the characteristic polynomial, either the built-in **charpoly** command or the determinant of the matrix A - xI. Substitution of the matrix A for the variable x in the characteristic polynomial requires the **evalm** command. The result is the zero matrix, indicating that A, indeed, satisfies its characteristic equation.

There are several ways to find the minimal polynomial, the polynomial p(x) of least degree for which p(A) = 0. Since the minimal polynomial must divide the characteristic polynomial, and since the characteristic polynomial here is just $(1 - x)^4$, we have only a few options for the minimal polynomial.

```
> mp := factor(minpoly(A, x));
> for k from 1 to 3 do evalm(subs(x = A, (1 - x)^k)); od;
```

Since A is not the identity, we could have predicted that k = 1 would not lead to the minimum polynomial. In either event, we have determined the minimal polynomial to be $(1 - x)^2$.

The Jordan form for A must therefore have at most a 2x2 block, and at least one such block. The possibilities are then two 2x2 blocks or one 2x2 block and two 1x1 blocks. In the first case the eigenspace for the repeated eigenvalue 1 will be two-dimensional, but it will be three-dimensional in the second case.

```
> J := jordan(A);
> jordan(A, p);
> P := inverse(p);
> q := eigenvects(A);
```

Maple's **jordan** command has returned the Jordan canonical form for A. We immediately see that there are two 2x2 blocks corresponding to the eigenvalue $\lambda = 1$ whose algebraic multiplicity is 4. We also see that the geometric multiplicity is 2 since there is just one eigenvector associated with each of the 2x2 Jordan blocks in J. The alternate form of the **jordan** command includes a variable into which Maple writes the inverse of the matrix of eigenvectors and generalized eigenvectors corresponding to the entries in J. Hence, the matrix P is returned as the transition matrix for the similarity transformation that puts A into Jordan form via $J = PAP^{-1}$. From the result of the **eigenvects** command which returns the eigenvalue $\lambda = 1$, its algebraic multiplicity of 4, and just two eigenvectors, we see that the first column of J is a multiple of the first eigenvector, and the third column of J is the second eigenvector.

To construct the transition matrix P, we can identify two generalized eigenvectors, one terminating each chain of two vectors, **v2** and **w2**, and work back to the two eigenvectors **v1** and **w1** found by the **eigenvects** command.

```
> v2 := vector([0,0,1,0]);
```

```
> w2 := vector([1,0,0,0]);
> v1 := evalm((A - 1)*v2);
> w1 := evalm((A - 1)*w2);
> S := augment(w1, w2, v1, v2);
> evalm(inverse(S) &* A &* S);
```

The transition matrix S so constructed is the same as P, the one Maple delivered.

Example 5. Repeat the analysis of Example 4 with the more complicated matrix

$$A = \begin{pmatrix} 10 & -4 & -9 & 0 & 0 & 9 \\ -3 & 14 & 4 & -5 & -4 & 1 \\ 4 & -10 & -3 & 4 & 3 & 1 \\ -11 & -16 & 11 & 5 & 9 & -12 \\ -5 & 26 & 7 & -11 & -7 & 4 \\ -8 & -11 & 8 & 4 & 6 & -9 \end{pmatrix}.$$

```
> r1 := [10,-4,-9,0,0,9]:
> r2 := [-3,14,4,-5,-4,1]:
> r3 := [4,-10,-3,4,3,1]:
> r4 := [-11,-16,11,5,9,-12]:
> r5 := [-5,26,7,-11,-7,4]:
> r6 := [-8,-11,8,4,6,-9]:
> A := matrix([r.(1..6)]);
```

We have entered the matrix A by rows in the hope that this will be less error prone than entering A through a continuous list of 36 elements. We next obtain the characteristic polynomial, show that A satisfies its characteristic equation, find the minimum polynomial, the eigenvalues and eigenvectors of A, and its Jordan canonical form with transition matrix P.

```
> cp := factor(charpoly(A, x));
> evalm(subs(x = A, cp));
> mp := factor(minpoly(A, x));
> q := eigenvects(A);
> J := jordan(A, p);
> P := inverse(p);
```

From these calculations we see that A has $\lambda = 1$ as an eigenvalue with algebraic multiplicity 2 but geometric multiplicity 1, and has $\lambda = 2$ as an eigenvalue with algebraic multiplicity 4 but geometric multiplicity 2. Hence, the Jordan form contains a 2x2 block for the eigenvalue $\lambda = 1$ and two blocks related the eigenvalue $\lambda = 2$. How the associated 4x4 region in J is arranged would not as yet be clear were it not for the information in the minimum polynomial. The lengths of the chains for each of the eigenvectors of $\lambda = 2$ can be seen from J to be 3 and 1, but that is known by observing J. These chain lengths are the powers on the factors of the minimum polynomial. But there is little difference in obtaining that information from Maple via the minimum polynomial or via the actual Jordan form itself.

For the chain associated with the eigenvector for $\lambda = 1$, we begin by looking at the null spaces of $(A - I)^k$ for $k = 1$ and 2.

```
> N1 := nullspace(A - 1);
> N2 := nullspace(((A - 1)^2);
```

Let **v2** be a vector in the null space N2 that is not in the null space N1, and hence, is not itself an eigenvector belonging to $\lambda = 1$. This is the generalized eigenvector at the end of the chain belonging to **v1**, the eigenvector for $\lambda = 1$, which we compute next. Great care must be exercised when extracting **v2** from the set N2. Different executions of the **nullspace** command can return the basis vectors in different orders. To make matters worse, it can even happen that use of the selector notation N2[1] will result in the second vector in the display being chosen! Only observation, care, and careful testing will guarantee selecting the vector in N2 that is not in N1.

```
> v2 := N2[2];
> v1 := evalm((A - 1)*v2);
```

Next, consider the null spaces for $(A - 2I)^k$, $k = 1, 2, 3$. From the minimal polynomial we know there is a chain of length 3, so there must be a chain of length 1 to account for the algebraic multiplicity of 4. The chain of length 1 is just a single eigenvector, which we already have. The generalized eigenvector at the end of the chain of length 3 is the more challenging target.

```
> NN1 := nullspace((A - 2));
> NN2 := nullspace((A - 2)^2);
> NN3 := nullspace((A - 2)^3);
```

Pick **w3**, the chain-terminating generalized eigenvector from the null space NN3, being sure it is not in NN2 or NN1. We select **w3** = (1,0,1,0,0,0), and either select it from NN3 or enter it directly.

```
> w3 := NN3[3];
```

The rest of the chain (**w2** and **w1**) is computed as follows.

```
> w2 := evalm((A - 2)*w3);
> w1 := evalm((A - 2)*w2);
```

The chain that consists of the single eigenvector can be either member of the null space NN1. In fact, only one of the two members of NN1 is independent of **w1**. We'll pick **u1** to be NN1[1].

> rank(augment(w1, NN1[1]));
> rank(augment(w1, NN1[2]));
> u1 := NN1[1];

Form the transition matrix S from **v1**, **v2**, **w1**, **w2**, **w3**, **u1**, then test S to see if it indeed brings A into Jordan form.

> S := augment(v1, v2, w1, w2, w3, u1);
> evalm(inverse(S) &* A &* S);

We indeed obtain J, and therefore note that the transition matrix S is not unique. Both P, the matrix computed by Maple, and S, the matrix we directed Maple to form, generate similarity transformations which bring A into Jordan canonical form.

Question 3. For each of the following matrices, find the characteristic polynomial and verify that the matrix satisfies its characteristic equation. Determine the minimum polynomial, the eigenvalues and eigenvectors, the Jordan form, and the transition matrix P computed by Maple. Use the method of Example 5 to determine the Jordan form and the transition matrix S (which may indeed be other than P) and verify that *your* S also generates a similarity transformation bringing A into Jordan form.

$$(a) \begin{pmatrix} -3 & 4 & 1 & 3 \\ -7 & 7 & 2 & 4 \\ 1 & 0 & 1 & 0 \\ 1 & 0 & -1 & 2 \end{pmatrix}; (b) \begin{pmatrix} 0 & 2 & 1 & 1 \\ -5 & 6 & 2 & 2 \\ 2 & -1 & 1 & 0 \\ 2 & -1 & -1 & 2 \end{pmatrix}; (c) \begin{pmatrix} 0 & 1 & 1 & 0 \\ -4 & 4 & 2 & 1 \\ 2 & -1 & 0 & 0 \\ 3 & -2 & -1 & 0 \end{pmatrix}.$$

INDEX OF MAPLE COMMANDS

* Numbers reference the Labs

INDEX OF TOPICS

*Numbers reference the Labs